TRANSBOUNDARY WATER MANAGEMENT IN A CHANGING CLIMATE

PROCEEDINGS OF THE AMICE FINAL CONFERENCE, SEDAN, FRANCE, 13–15 MARCH 2013

Transboundary Water Management in a Changing Climate

Editors

Benjamin Dewals
University of Liege, Department ArGEnCo, HECE – Hydraulics in Environmental and Civil Engineering, Liege, Belgium

Maïté Fournier
EPAMA, Charleville-Mézières, France

CRC Press
Taylor & Francis Group
Boca Raton London New York Leiden

CRC Press is an imprint of the
Taylor & Francis Group, an **informa** business

A BALKEMA BOOK

Cover photos information:
Top left: The river Meuse in Commercy (France, photo: EPAMA)
Top right: Meander of the river Meuse (Fépin, France, photo: A. Collard)
Bottom left: Inundations in the floodplains of the river Meuse (Wallonia, Belgium, photo: SPW)
Bottom right: Low flows in the border Meuse (Belgium / the Netherlands, photo: M. Lejeune)

CRC Press/Balkema is an imprint of the Taylor & Francis Group, an informa business

© 2013 Taylor & Francis Group, London, UK

Typeset by MPS Limited, Chennai, India

Published by: CRC Press/Balkema
P.O. Box 11320, 2301 EH, Leiden, The Netherlands
e-mail: Pub.NL@taylorandfrancis.com
www.crcpress.com – www.taylorandfrancis.com

ISBN: 978-1-138-00039-1 (Hbk)
ISBN: 978-0-203-74447-5 (eBook)

Transboundary Water Management in a Changing Climate – Dewals & Fournier (Eds)
© 2013 Taylor & Francis Group, London, ISBN 978-1-138-00039-1

Table of contents

Map of the Meuse basin

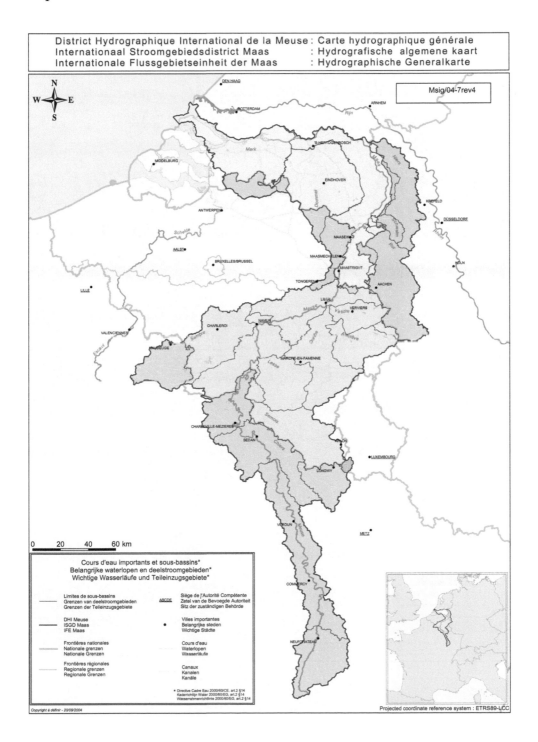

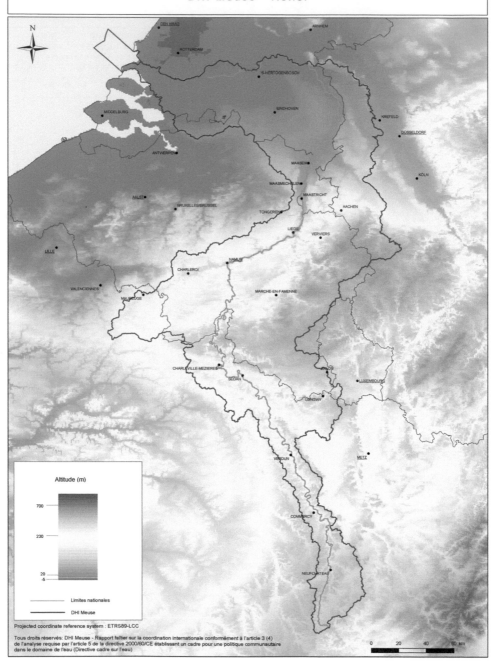

DHI Meuse - Relief

N

DEN HAAG
ARNHEM
ROTTERDAM
'S-HERTOGENBOSCH
EINDHOVEN
KREFELD
MIDDELBURG
DÜSSELDORF
ANTWERPEN
MAASEIK
KÖLN
MAASMECHELEN
AALST
MAASTRICHT
BRUXELLES/BRUSSEL
AACHEN
TONGEREN
LIEGE
LILLE
VERVIERS
NAMUR
CHARLEROI
VALENCIENNES
MARCHE-EN-FAMENNE
MAUBEUGE

CHARLEVILLE-MEZIERES
ARLON
SEDAN
LUXEMBOURG
LONGWY

VERDUN
METZ

COMMERCY

NEUFCHATEAU

Altitude (m)

700
230
20
-5

—— Limites nationales
—— DHI Meuse

Projected coordinate reference system : ETRS89-LCC

0   20   40   60 km

Foreword by EPAMA's President

J.P. Bachy
EPAMA, Charleville-Mézières, France

Dear Sir, Dear Madam,

After 5 years of rich and varied activities on the Meuse river basin, the AMICE Project is nearly completed. As the Lead Partner, I have the pleasure to invite to our final conference and give us the pleasure to celebrate the success of this international cooperation.

You will have the opportunity to meet all the Partners that were deeply involved in this ambitious adventure. In addition to the presentation of AMICE's achievements, we have invited experts who will tell us about other innovative approaches for water management and adaptation to climate change.

Finally, the animation programme and the workshops will offer us the possibility to proceed with our collaboration and to think about the future. The Meuse river is a transnational structuring line, that shall be regarded as a development leverage intended to the federation of local, regional and international stakeholders.

Looking forward to welcoming you at this event!

Madame, Monsieur,

Après 5 années d'activités riches et variées sur le bassin versant de la Meuse, le Projet AMICE touche à sa fin. En tant que Chef de File, je souhaite vous convier à notre conférence de clôture qui sera l'occasion de célébrer la réussite de cette coopération internationale.

Vous aurez l'occasion d'y rencontrer tous les Partenaires qui se sont pleinement investis dans cette ambitieuse aventure. En plus de la présentation des résultats d'AMICE, nous avons invité des experts qui nous feront découvrir d'autres approches innovantes pour la gestion de l'eau et l'adaptation au changement climatique.

Enfin, le programme d'animation et les ateliers de réflexion nous donnerons l'opportunité de poursuivre notre collaboration et d'envisager l'avenir. La Meuse est un axe structurant transfrontalier, qui doit être aussi considéré comme un levier de développement, qui a vocation à fédérer tous les acteurs locaux, régionaux et internationaux concernés.

Dans l'attente de vous accueillir lors de cet évènement!

Transboundary Water Management in a Changing Climate – Dewals & Fournier (Eds)
© 2013 Taylor & Francis Group, London, ISBN 978-1-138-00039-1

Programme and Speakers

Wednesday March 13th, 2013
Galerie des Antiques, Château Fort, Sedan

18:30	Welcome Speeches	Conseil Régional de Champagne-Ardenne
		Conseil Général des Ardennes
		Préfecture des Ardennes
		Agence de l'Eau Rhin-Meuse
19:30	Ice-Breaking Party	

Thursday March 14th, 2013
Salon Prestige, Stade Louis Dugauguez, Boulevard de Lattre de Tassigny, Sedan

09:30	Opening Live-Act		
10:00	Plenary session 1:	Harry Tolkamp	Waterboard Roer en Overmaas
	DOES THE RIVER	Patrick Willems	Catholic University of Louvain
	MEUSE CHANGE?	Benjamin Dewals &	University of Liège &
		Gilles Drogue	University of Lorraine
10:50	Coffee break		
11:20	Plenary session 2:	Paul Dewil	Walloon Region (GTI)
	TOO MUCH OR	Benjamin Sinaba	University of Aachen
	TOO LITTLE	Harry Romgens	RIWA Maas
	WATER	Martine Lejeune & Carole	RIOU & Community of Hotton &
		Raskin & Piet Van Iersel	Waterboard Brabantse Delta
12:25	Lunch break		
14:05	Plenary session 3:	Frank Mostaert	Flanders Hydraulics Research
	WHAT ARE WE	Joop DeBijl & Koen	Waterboard Aa en Maas &
	DOING?	Maeghe & Gerd Demny	nv DeScheepvaart &
			Wasserverband Eifel-Rur
		Bruno Tonnelier	Etat Major de Zone – Est
		Bieri Martin	École Polytechnique Fédérale
			de Lausanne
15:10	Coffee break		
15:40	Plenary session 4:	Thomas Borchers	BMU.BUND
	LIVING WITH	Hendrik Buiteveld	Rijkswaterstaat
	CLIMATE CHANGE	Roberto Epple	European River Network
16:30	Closing session	Maïté Fournier	EPAMA
17:00	End		
18:00	Guided tour	Medieval Castle in Sedan	
		Ancient city of Sedan	
		Brewery of the Castle in Sedan and beer tasting	
		Bus tour "land of memory" around Sedan	

Château du Faucon, Rue de l'atelier, Donchery, Sedan

19:30	Conference Diner

Friday March 15th, 2013
Salon Prestige, Stade Louis Daugauguez, Boulevard de Lattre de Tassigny, Sedan

09:30	Plenary session 5: TRANSNATIONAL WATER MANAGEMENT	Professor van Ypersele Liina Tuulik	IPCC Interreg IV B Secretariat
	COMES TRUE	Maïté Fournier	EPAMA
10:45	Coffee break		
11:15	Workshop 1: THE MANY FACES OF THE MEUSE	Frederiek Sperna Weiland Heribert Nacken	Deltares University of Aachen
11:15	Workshop 2: THE WHIMSICAL MEUSE	Colin Green Aurore Degré	CONHAZ project University of Liège – Gembloux AgroBioTech
11:15	Workshop 3: TAMING THE MEUSE ?	Régis Thépot Max Linsen	EPTB Seine Grands Lacs Rijkswaterstaat
11:15	Workshop 4: WE AND THE MEUSE	Evelyne Huyghe Martine Lejeune	Future Cities project RIOU
11:15	Workshop 5: A VISION FOR TOMORROW ON THE MEUSE BASIN	Secretary General Sonja Koeppel	International Meuse Commission UNECE
12:15	Press conference		
12:30	Lunch break		
14:15	Conclusions of the Workshops	Xavier Caron	EPAMA
14:45	MEUSE COUNTRIES' JOINT DISCUSSION	*With representatives from:* Conseil Régional de Lorraine Walloon Region Flemish Region Association of Dutch Waterboards Federal representation of the Lander North-Rhine Wesfalia	
15:45	Closing speech	Jean-Paul Bachy	President of EPAMA
16:00	End of conference		

Programme on Novembre 30th, 2012

Transboundary Water Management in a Changing Climate – Dewals & Fournier (Eds)
© 2013 Taylor & Francis Group, London, ISBN 978-1-138-00039-1

Adaptation of the Meuse to the impacts of climate evolutions – the AMICE project in short

M. Fournier
EPAMA, Charleville-Mézières, France

M. Lejeune
RIOU, Hasselt, Belgium

ABSTRACT: Climate evolutions impact the Meuse basin creating more floods and more droughts. The river managers and water experts from 4 countries of the basin join forces in this EU-funded transnational project to elaborate an innovative and sustainable adaptation strategy. The project runs from 2009 through 2013.

1 OBJECTIVES

– Defining a common strategy of adaptation to the impacts of Climate Change on floods and droughts, recognized at the scale of the international river basin of the Meuse. Climate scenarios for the time periods 2020 2050 and 2070–2100, existing measures, on-going projects as well as the Floods Directive (2007/60/EC), will be taken into account in the elaboration of the strategy.
– Realizing a set of measures beneficial and transferable to the whole Meuse basin.
– Strengthening and widening the partnership of stakeholders in the international Meuse basin.
– Involving the population and the public bodies through a better knowledge and the feeling of belonging to the Meuse basin, as well as the consciousness of flood and drought risks.

Adaptation of the Meuse to the Impacts of Climate Evolutions

2 ORGANIZATION AND EXPECTED OUTPUTS

– Definition of scenarios, shared at the scale of the international basin, related to climate change and extreme discharges;
– Realization of a first hydraulic simulation of the river and its associated risk maps;
– Identification of hot spots, i.e. sectors and water uses threatened by future floods and droughts;
– Definition of a shared adaptation strategy;

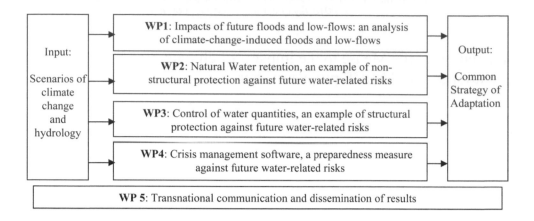

- Six technical reports, one interactive documentary and a website, nine site visits;
- Definition of new management rules for the dams of the Rur, tributary of the Meuse (Germany);
- Reduction of the floodability, renaturation of cropland and creation of reservoirs for multi-purposes (Wallonia and the Netherlands);
- An innovative system of pumps and powerstation for the reduction of water consumption in the Meuse (Flanders);
- An international exercise on flood crisis management based on the software OSIRIS and FLIWAS;
- An international event for the sharing and transfer of the project's results.

3 PARTNERSHIP

The AMICE project involves 17 partners from the Meuse basin. The International Meuse Commission hosts the Partners' meetings of the AMICE project and acts as an observer.

3.1 *France*

- EPAMA (Public Organisation for the Management of the Meuse and its Tributaries), also Lead Partner of the AMICE project;
- Université de Metz – department CEGUM (Centre for Geographical studies);
- CETMEF (Institute for Maritime and Inland Waterways).

3.2 *Belgium – Wallonia*

- Région Wallonne, through the cross-disciplinary working group on floods (GTI);
- Université de Liège – Department of Hydraulics in Environmental and Civil Engineering;
- ULg, Gembloux Agro-Bio Tech – department Hydrology and Hydraulics;
- Municipality of Hotton;
- Agence Prévention et Sécurité (APS).

3.3 *Belgium – Flanders*

- nv De Scheepvaart, manager of the channels for water transport and drink water production;
- Waterbouwkundig Laboratorium, the research center for hydraulic sciences in Antwerp;
- RIOU asbl, association for communication and renaturation.

3.4 *Germany*

– WasserVerband Eifel-Rur, manager of the Rur tributary;
– RWTH Aachen Universität: Lehrstuhl und Institut für Wasserbau und Wasserwirtschaft: the institute of hydraulic engineering and water resources management;
– RWTH Aachen Universität: Lehr- und Forschungsgebiet Ingenieurhydrologie: the academic and research department engineering hydrology.

3.5 *The Netherlands*

– Rijkswaterstaat is involved through two of its departments: Waterdienst and Limburg;
– Waterschap Aa en Maas;
– Waterschap Brabantse Delta.

4 BUDGET

The AMICE application form was approved on the 3rd call for projects of the INTERREG IV B Programme on the 5th of December 2008; 2.8 M€ ERDF were granted to the project. The total budget of the AMICE project is nearly 8.9 M€.

5 THE NWE INTERREG IV B PROGRAMME

The Programme funds innovative transnational actions that lead to a better management of natural resources and risks, to the improvement of means of communication and to the reinforcement of communities in North-West Europe.

REFERENCES

www.amice-project.eu
www.amice-film.eu
www.nweurope.eu

Transboundary Water Management in a Changing Climate – Dewals & Fournier (Eds)
© 2013 Taylor & Francis Group, London, ISBN 978-1-138-00039-1

Does the river Meuse change?

H.H. Tolkamp
Roer and Overmaas Regional Water Authority, Sittard, The Netherlands,
Project Manager of FLOOD-WISE

ABSTRACT: Flood risk management in cross-border areas can benefit from international river basin commissions that can set up and carry out river basin management directed at flood and water quality issues across borders. Thus single sided measures can be avoided and opportunities to harmonize cost recovery, spatial planning, public participation and communication pave the way to real cross border cooperation. The risk of floods, both the frequency as the gravity, is increasing with the gradual change of the climate. Transboundary governance and river basin wide management can contribute to damage prevention, reduction and mitigation of the negative effects of inundations. It is essential that experts and managers keep exchanging their experiences and knowledge of the river system.

1   INTRODUCTION

River basin management and flood risk management is not the business of the water authorities that happens to be responsible for the water management along the river. It is the collective responsibility of all water authorities in the river basin, from source to mouth. This is in fact the basic principle under the European Water Framework Directive and the Floods Directive. Do adapt and cooperate with all authorities in a river basin and do not transport your problems down river, of take action that cause problems down stream. Cooperate and agree with all authorities as if there were no borders between nations or administrations in the river basin.

2   ROER AND OVERMAAS REGIONAL WATER AUTHORITY

To be given the opportunity to introduce the two invited speakers on the subject of the influence of climate change on river dynamics is a challenge. One should not tell the plot of what is coming, but one should try to get everyone's attention.

Introducing the nature and the necessity of river dynamics to keep rivers alive, would take a special course during a whole semester probably. But making the audience aware of the fact that floods have been of all times and nowadays do depend largely on the combination of natural phenomena and the human preparedness is something that takes less time. This will be the subject of the conference these two days.

I am known to some of you as an old hand working on biological assessment with the aid of macro-invertebrates, and to some of you as the stimulator of river and stream restoration.

That is indeed where my working contribution to water management started in the early seventies, and I can say that this has been situated in an international context from the beginning and this has continued for a few decades. I also was involved in the work of the first Commission for the Protection of the River Meuse as chair of the sub-working group for the inventory of diffuse pollution and since then I have been involved in the work of the International Meuse Commission, advising on several subjects.

The same goes for the work of the Bilateral Flemish-Dutch Meuse Commission. This is part of my job as a senior strategic policy advisor of the Roer and Overmaas Regional Water Authority where I have been active in the international water management since 1980. This is really not a coincidence, since this Water Authority is the most international one of the 25 Water Authorities working in the Netherlands in 2012.

Roer and Overmaas was involved in the conception of the AMICE project, but did not become a direct partner because our neighbouring water managers (Rijkswaterstaat Limburg and WVER) could very represent us and involved us where needed.

3 PROJECT FLOOD-WISE

Another direct involvement with the River Meuse follows from my part time job in the last two years as Project Manager of the INTERREG IV C project FLOOD-WISE.

A project that ran for three years and that was concluded just a months ago, working on gaining and exchanging experience, knowledge and good practices between 10 countries (7 EU and 3 non-EU countries) on the transboundary implementation of the EU Floods-Directive. A project especially aimed at communication, sharing, learning, cooperating, which three years of interaction and pilot-projects on six cross-border river basins (Figure 1). Working on sustainable flood risk assessment, flood hazard and flood risk mapping and gaining experiences in flood risk management.

A project that led to an incredible positive cross-border cooperation and the common opinion now the project funding has ended, that the network that was formed should be kept. Therefore, we have proposed to form an umbrella to protect the existing network and to give others the opportunity to join. As you know the best example of a real life network, illustrating the complexity and vulnerability of a network, is the spiders web, a clever and complicating construction that is rebuilt every day when necessary. That is something we cannot afford, because we cannot risk missing important knowledge that is passing when the web is in reconstruction. We need to keep the network alive and functioning to catch our ideas and our dreams to be able to share these thoughts. This kind of network is not unique for FLOOD-WISE and is true for AMICE just the same.

So please get acquainted with the Task Force Water Management for cross-border areas and get involved, so we can come up with new ideas and new projects to further our cooperation within Europe. The Task Force aims at keeping the knowledge and expertise together and concentrated, to be able to continue existing partnerships and enhance new partnerships; to take initiative for new cross border applications and projects and to organize meeting opportunities on regular basis (www.floodwise.nl; TFWaterGovernance@gmail.com). This must function as a breeding pond for project ideas, which can be forged into project proposals by combining our knowledge and strength.

FLOOD-WISE and AMICE are both INTERREG projects aiming at the improvement of cross border cooperation, working on similar problems and with the Meuse as a common denominator.

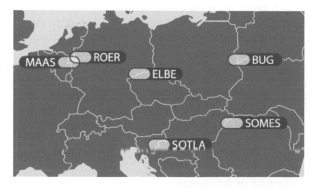

Figure 1.   Pilot-projects were conducted on six cross-border river basins.

It would be very wise to add the knowledge and experience gained during the projects life to the Task Force, to the network. This is already supported by several European organizations:

- Eurisy in Paris (NGO from European Space Agency) aiming at how to apply satellites knowledge and data in water management; BENELUX in Paris working on how to deal with cross border structures in water management (European Grouping for Territorial Cooperation);
- the Association of European Border Regions, where the Task Force will have a prominent place on the agenda and will take part in the next AEBR annual meeting in Liege.

4 DOES THE RIVER MEUSE CHANGE?

So does the River Meuse change? Let's certainly hope so, because if she didn't the Meuse would have lost its characteristics of a river. The basic character of a river is the natural and constant dynamic, including the variations in current velocity, discharge volume, river bed and bank erosion, the whole package of the dynamics that make a river what it is. And now we know for certain that our climate is changing, we can be sure that the Meuse will adapt to these changes.

5 CONCLUSION FROM FLOOD-WISE

It would be very helpful to the European countries if the Floods Directive would be supported by a catalogue of potential objectives and related measures, including experiences and good practices already available. The WISE-RTD web portal (www.wise-rtd.info) provides an excellent tool to support this approach, facilitating the exchange and retrieval of good practices.

Transboundary river basins will benefit from international river basin commissions to set up and carry out integrated river basin management directed at flood and water quality issues across borders. Through these commissions single sided measures can be avoided, and opportunities to harmonize cost recovery, spatial planning, public participation and communication pave the way to real cross border cooperation.

Measures that include spatial planning restrictions should be enforced with priority, because integrated flood risk management depends on all links of the chain to be in place and equally strong.

6 THIS SESSION IN THE FINAL CONFERENCE OF AMICE

In the next contributions, two experts will take us by the hand and show us what is happening.

Patrick Willems of the CCI-HYDR will introduce the subject. He is working at the Catholic University of Leuven on climate change as the coordinator of the CCI-HYDR project (Climate change impact on hydrological extremes along rivers and urban drainage systems in Belgium) and we look forward to his views on the future development of the effects of Climate Change on the system of the river Meuse.

After his presentation, Benjamin Dewals of the University of Liege and Gilles Drogue of the University of Metz will reflect on these subjects from the back ground and experiences gained in the project AMICE. Both Benjamin and Gilles are involved in the AMICE project from the beginning and they have been concerned with climate change and hydrological impacts for many years.

Transboundary Water Management in a Changing Climate – Dewals & Fournier (Eds)
© 2013 Taylor & Francis Group, London, ISBN 978-1-138-00039-1

Multidecadal climate oscillations and climate scenarios for impact analysis on hydrological extremes in Belgium

P. Willems
KU Leuven, Leuven, Belgium

ABSTRACT: Climate variability, climate change and the impact on hydrological extremes have been studied for Belgium. The analysis of the climate variability was based on a more than 100-year time series of 10-minute rainfall intensities at Uccle, Brussels. Also comparison was made with other hydro-climatic series of neighbouring regions in Western Europe. Future climate change was studied based on a set of climate scenarios that were tailor-made for impact analysis on hydrological extremes in Belgium. They were based on an ensemble set of more than 50 global and regional climate model simulations and after statistical analysis, downscaling and intrinsic bias correction using an advanced quantile perturbation approach. The scenarios were applied for the analysis of various types of hydrological and related impacts in Belgium. These include impacts on river peak and low flows, river floods, surface water availability, groundwater, urban drainage, urban floods, coastal floods, vegetation and ecology, crop growth and agriculture, etc. This paper gives an overview.

1 INTRODUCTION

Flood risk is in Belgium as well as in other European countries of considerable importance. This is due to dense populations, high level of urbanization and high industrialization. Since last decades, sewer systems are being built at a large scale. Drought risks are less significant in the country, due to the humid climate and the limited length of the dry spells in summer. However, extreme low flows may occur along rivers, causing severe problems of water shortage for drinking water supply, for agriculture and for the environment.

There is strong evidence that due to global change, the risks of inundations, low flows and water availability are changing. The consequences of these changes in potential hazards are to be assessed in a perspective of sustainable development. Authorities, water suppliers and water dependent sectors have to anticipate these changes, as to limit the flood and water availability risks of the inhabitants to acceptable risk levels.

The concerns about the impact of climate change on the hydrological water cycle have triggered specific studies since the 1980s. The Royal Meteorological Institute of Belgium (RMI) has been pioneering in putting into evidence differences in the sensitivity of catchments with contrasted characteristics to a $2 \times CO_2$ scenario (e.g., Bultot et al., 1988). They extended their study to a larger set of catchments and used the first set of climate change scenarios made available by the IPCC (Gellens & Roulin, 1998). The scope was further extended to the whole Meuse river basin (Roulin et al., 2001) and Scheldt river basin (Roulin & Arboleda, 2002) using a new set of climate scenarios based on transient experiments, for instance based on the results of Global Circulation Models (GCMs) forced with an increasing greenhouse gas content. However, the GCMs have since improved, and high resolution regional climate models (RCMs) have been nested within to downscale the climate variables to regional scale. This has sparked new research related to regional impacts relevant at local scales. Hydrological impact assessments can now be performed with increased confidence.

The CCI-HYDR research project, funded by the Belgian Science Policy Office (BelSPO), and carried out by the Hydraulics division of KU Leuven (prof. Patrick Willems) and the Royal Meteorological Institute of Belgium (ir. Emmanuel Roulin, dr. Pierre Baguis), investigated in a detailed objective way and based on the most recent data and climate modelling results, the climate change impact on the risk of hydrological extremes along rivers and urban drainage systems in Belgium (Willems et al., 2010). After a study of the climate variability during the past 100 years at Uccle, Brussels, a large ensemble set of available simulation results from GCMs and RCMs were analysed and downscaled to the time and space scales required for hydrological impact analysis. The scenarios were applied for impact modelling towards flood risks and low flows risk along rivers, and flood risks along urban drainage systems, making use of hydrological and hydrodynamic models. Afterwards the CCI-HYDR climate scenarios were extended for impact analysis along the Belgian coast, on agriculture and certain types of environmental impact assessment. The scenarios are in the meantime also applied by many others for various types of water and wet/dry climate conditions related impacts.

2 DETECTION OF MULTIDECADAL CLIMATE OSCILLATIONS

Given the focus of the project on hydrological extremes, long term climate variability has been studied by analysing changes in extreme rainfall quantiles. This has been done by means of the full time series of 10-minute rainfall intensities at Uccle, Brussels, since 1898, collected and processed by the RMI. From that time series, independent rainfall extremes were extracted at various time scales ranging from 10 minutes to 1 month. Based on a technique for the identification and analysis of changes in extreme quantiles, it is shown that hydrological extremes have oscillatory behaviour at multidecadal time scales (Ntegeka & Willems, 2008a).

The past 100 years showed increases in the magnitude and frequency of rainfall extremes around the 1910s, 1950–1960s, and more recently during the 1990–2000s. Similar oscillations were found for river flow extremes in the Meuse river at Borgharen since 1925 (Willems & Yiou, 2010).

After application of the same method to more long rainfall series across Europe, similar oscillations were found for neighbouring stations. The results moreover showed that the oscillations for south-western Europe are anti-correlated with these of north-western Europe, thus with oscillation highs in the 1930–1940s and 1970s. The rainfall oscillations appear explained by persistence in atmospheric circulation patterns over the North Atlantic during periods of 10 to 15 years (Willems & Yiou, 2010).

Next to these oscillations, significant trends in rainfall extremes were detected during the most recent decades, but only for the winter season. This trend is likely the result of climate change; the trend is at least consistent with the range of climate impact projections made by climate models (see next section).

3 CLIMATE CHANGE SCENARIOS FOR IMPACT ANALYSIS ON HYDROLOGICAL EXTREMES IN BELGIUM

3.1 *Climate scenarios for impact analysis on wet–dry extremes along rivers and catchments*

Future climate scenarios were first derived for rainfall and potential evapotranspiration (ETo) based on a set of about 30 simulations derived from 10 RCMs nested in two main GCMs (Baguis et al., 2009). These are publicly available through the database of the EU PRUDENCE project. The experiments were run for the IPCC regional A2 and B2 future greenhouse gas emission scenarios. Since the PRUDENCE RCM models are based on only the A2 and B2 scenarios, scaling factors were applied to make the scenarios more exhaustive (thus better taking into account the GCM model uncertainty and the emissions uncertainty) by including changes from extra GCMs and extra emission scenarios (notably the A1B and B1 scenarios). For that purpose, all GCM simulations available for Belgium in the Archive of the 4th Assessment Report of the IPCC were considered.

The selected RCM simulations exhibited both negative and positive changes (-40% to $+10\%$) in rainfall during the hydrological summer, and positive changes during the hydrological winter ($+5\%$ to $+50\%$). There were no significant regional differences in the climate change signals over Belgium; with the exception of the coastal region. The rainfall increases are for the main Belgian lands around 10% lower than the ones over the coastal area.

From the large set of climate model projections, three probabilistic scenarios were extracted to allow end users to investigate the range of changes. The scenarios were appropriately named high/wet, mean/mild and low/dry. They represent scenarios that are expected to lead to highest/ median/lowest impacts on wet/dry conditions in winter/summer, hence tailor-made for impact analysis on hydrological extremes (Ntegeka et al., in revision). The scenarios take the seasonal dependency between the rainfall changes and the ETo changes into account, hence try to preserve the physical consistency in the climatic changes between seasons and variables.

In order to deal with the problem of time and space scale difference between the GCM and RCM results and the hydrological impact models, the bias in the GCM and RCM results (as observed for current climatic conditions), and to enable easy application of the climate scenarios in practice, a perturbation tool has been developed. This CCI-HYDR Perturbation Tool translates the climate scenarios to changes in the historical rainfall and ETo input series of hydrological models (Ntegeka & Willems, 2008b). This is done by an advanced version of the delta change method: the quantile perturbation approach. For the rainfall series, the approach involve both changes in the frequency of rain storms and changes in the rainfall intensity. The changes are being made in a variable way, depending on the month in the year, and as a continuous function of the return period or rain storm frequency. For the ETo series, only intensities are perturbed, but also depending on the month and the return period. The series to be perturbed can be daily or hourly and can have any length (typical lengths vary from a few years to 100 years). The perturbations can be made for time horizons till 2100 (e.g., for 2020, 2030, …, 2100).

3.2 Extension for ecological Impact analysis

The first CCI-HYDR scenarios were later extended to account for more variables, notably the ones of highest importance for environmental impact assessment: temperature and wind speed. Also a comparison was made with the KNMI'06 climate scenarios for The Netherlands. The extended CCI-HYDR climate scenarios were considered on the basis of the Flanders' Environmental Assessment 2030 by the Flemish Environment Agency (Vlaamse Milieumaatschappij: VMM) as well as the Flanders' Nature Assessment 2030 by the Flemish Institute for Nature and Forest Research (Instituut voor Natuur – en Bosonderzoek: INBO) (Brouwers et al., 2009; Willems et al., 2009).

3.3 Extension for impact analysis on urban drainage systems

Another extension of the CCI-HYDR scenarios was made for the for the Flemish Environment Agency (VMM) to allow impact investigation along urban drainage systems and other small scale storage facilities (storage tanks and basins). This required further downscaling to the 10-minute time scale, and a more thorough investigation of the statistical downscaling step. The latter was done by comparing several statistical downscaling methods, including methods based on weather typing or resampling. Interesting conclusion from that study was that the downscaling results of good methods do not differ much (Willems & Vrac, 2011).

The results were used to revise the Flanders' guidelines for the design of urban drainage systems, including new design storms for high-mean-low climate scenarios, this time tailor-made for urban hydrological impact analysis (Willems, 2009).

Recently, the scenarios were further extended by incorporating more climate model simulations, extending the set of PRUDENCE RCM runs to a larger ensemble set of RCM simulations obtained from the EU ENSEMBLES project. Also additional GCM runs were considered, bringing the total number of RCM runs for Belgium to 47 and the total number of GCM runs for Belgium to 69 (Willems, submitted).

11

3.4 *Extension for impact analysis along the Belgian coast and Scheldt Estuary*

Another extension was made to enable impact assessment of climate change along the Belgian Coast and the Scheldt Estuary. The inland climate scenarios for rainfall, temperature and ETo were for that purpose extended with corresponding scenarios for mean sea level, storm surge levels and coastal wind climate (direction and velocity). Correlations were studied between the coastal scenarios and the inland scenarios. These correlations are due to some atmospheric conditions that cause both extreme inland rainfall and high coastal surge levels. Changes in sea level pressure (SLP) were studied over the North Atlantic region and transferred to changes in storm surges at the Scheldt mouth (at Vlissingen). This was done based on a correlation model between the SLP at the Baltic Sea and the storm surge level. This model was derived after analysis of SLP composite maps and SLP-surge correlation maps for days where the surge exceeds given thresholds (for different return periods) (Ntegeka et al., 2012).

4 IMPACT OF CLIMATE CHANGE ON HYDROLOGICAL EXTREMES IN BELGIUM

4.1 *Impacts on river peak and low flows*

Hydrological impacts of the rainfall and ETo change scenarios were investigated at large scale in the Meuse and Scheldt basins, based on the SCHEME hydrological model of RMI (Baguis et al., 2010) and for 67 sub catchments in the Scheldt basin, based on the combined simulation of the NAM lumped conceptual rainfall-runoff models and the MIKE11 full hydrodynamic river model (Boukhris et al., 2008). In the latter model, floodplains were implemented based on a quasi-2D approach.

More detailed investigations were made for the catchment of the rivers Grote Nete – Grote Laak. For this catchment, different (types of) hydrological models were tested in order to study the uncertainty in the climate change impact results explained by the impact model. Six hydrological models of three different types were applied: the lumped conceptual rainfall-runoff models NAM, PDM and VHM, the detailed physically based and fully spatially distributed model MIKE-SHE, and the semi-distributed model WetSpa (Vansteenkiste et al., 2012, in revision).

Climate change impacts on hydrological extremes (floods and low flows) (scenario period 2071–2100 versus control period 1961–1990) indicated that this impact weakly depends on the topographical and soil type characteristics of the catchments. In general, low flows significantly decrease in all studied catchments and reaches up to 60% or more reduction in the low scenario. The increase in hourly river peak flow extremes is less strong, and limited to around 35%. Results indicate that low flow or drought problems will increase and might become more severe in comparison with flood risk problems induced by extreme precipitation.

Uncertainties in the results are, however, still very high. Depending on the ratio between the increase in rainfall versus the increase in ETo, and the ratio between the increase in winter rainfall versus the decrease in summer rainfall, the hydrological impact results for high flows might turn over from a positive trend into a negative trend.

While the climate change impacts tend towards wetter winters and drier summers, the hydrological response appears similar throughout the entire Scheldt and Meuse basin areas. The findings show that the intensity of the impacts is only slightly dependent on the location.

For the Grote Nete – Grote Laak catchment, the five models responded in a similar way to the future climate scenarios for the impacts on the peak flows. Only the PDM model simulated significantly higher runoff peak flows under future wet scenarios (up to 20% more by 2100), which is explained by its specific model structure. The hydrological model related uncertainties were higher for the impact results on low flows. The MIKE SHE model estimated the expected low flow decreases to be less severe under these scenarios with on average a lower decrease of about 30%. This was explained by its more detailed physically-based process descriptions for the groundwater and the groundwater-river interactions of the MIKE SHE model. The groundwater heads in the MIKE SHE model varied in the winter season between a few centimetres by 2100 in the lower, river valleys up to 1 m in the interfluvial and elevated zones. Summer heads will generally stagnate in the valleys to decrease up to 1 m in the higher zones. Results of all models moreover suggest that the model structure is more critical in low flow impact results of climate scenarios than in high flow conditions. The communication of these results has led to an increase of the awareness on future low flow and water availability problems in Belgium. That decrease in summer water availability may have severe consequences becomes clear if one calculates the mean water availability, which is about 1480 m^3/person/year for Flanders and Brussels and among the lowest in Europe due to the high population density and the high dependency on neighbouring regions for river inflow and outflow. Current groundwater abstractions are already non-sustainable and need urgent replacement by surface water abstraction, but future surface water limitations and water quality problems (reduction in dilution of pollution) in summer may hamper this change. Hence, there is a need to study adaptation strategies reg. low flow extremes and future water availability, taking uncertainties in the impact results into account.

4.2 *Impacts on urban drainage systems*

The impacts along urban drainage systems were quantified by means of a reservoir type of continuous simulation model. The changes in flood frequencies of sewer systems and overflow frequencies of storage facilities have been quantified. Also the change in storage capacity, required to exceed a given overflow return period, has been calculated, for a range of return periods and infiltration or through flow rates.

It was found that systems designed for a 2 years return period of flooding would flood twice that frequent for the high climate scenario (scenario period 2071–2100 versus control period 1961–1990). Regarding the design of local source control measures (storage facilities, rainwater tanks, infiltration reservoirs, etc), 15% to 35% increase in the storage capacity would be needed for the same climate scenario if one wants to limit the overflow frequency of the facility to the current level. Correspondingly, storage facilities with a current overflow return period of 2 years would overflow approx. twice per year; facilities with an overflow return period of 5 years would (for the same scenario) overflow approx. once per 1–1.5 years. The latter results indicate that there is a need for more and larger local storm water storage. In case this storage is built by means of infiltration ponds, the storm water stored in the ponds will enhance the groundwater infiltration and consequently will help to solve the enhanced low flow problems expected for river catchments.

Based on these results, changes were made to the Flanders' guidelines for the design of urban drainage systems. The design return period increased from 5 to 20 years, and more attention is given to source control: upstream infiltration on both public and private domains.

Inter-comparison was moreover made with the results and approaches by colleagues from other countries (Willems et al., 2012a).

4.3 *Impacts on groundwater*

More detailed investigation of the climate change scenarios on the groundwater system was made by Dams et al. (2012) for the Kleine Nete catchment. Instead of the high-mean-low climate scenarios, all PRUDENCE RCM based scenarios were simulated in a coupled WetSpa – MODFLOW model. The results show for the scenario period 2070–2100 compared with the reference period 1960–1990 a change in the annual groundwater recharge between −20% and +7%. On average annual groundwater recharge decreased 7%. In most scenarios the recharge increased during winter but decreased during summer. The altered recharge patterns caused the groundwater level to decrease significantly from September to January. On average the groundwater level decreased about 7 cm with a standard deviation between the climate scenarios of 5 cm. Groundwater levels in interfluves and upstream areas were more sensitive to climate change than groundwater levels in the river valley. Groundwater discharge is expected to decrease during late summer and autumn as much as 10%, though the discharge remains at reference-period level during winter and early spring.

4.4 Impacts on agriculture and related changes in catchment hydrology

Vanuytrecht et al. (2011, 2012) studied the impact on crop growth in Belgian agriculture as a result of the climate scenarios and related changes in soil moisture conditions and elevated CO_2 concentrations. This was done at field or parcel scale using the AquaCrop model. The climate scenarios had to be extended with the corresponding changes in CO_2 concentrations, which form the basis of the greenhouse gas emission scenarios used to force the different RCM and GCM runs. Changes in crop yield production as high as 27% of the total production were found, and +23% for the biomass production. This is because crops benefit from elevated CO_2 concentration by improving water productivity. +15% is achieved through production increases in aboveground biomass, +16% through increase in yield, and −5% by a decrease in seasonal evapotranspiration. Less critical, yet statistically significant are changes in canopy development rate and in phenology. In a next phase, it will be studied whether the changes in agricultural crop conditions in a catchment will change rainfall-runoff responses at catchment scale.

4.5 Impacts of correlated coastal-inland changes on river Scheldt

The impact of the correlated inland-coastal climate scenarios for rainfall, ETo, mean sea level, storm surge level and wind climate on the flood risk were studied for a hot spot area along the river Scheldt at Dendermonde (Ntegeka et al., 2012). This was done within the scope of the EU project Theseus. The Dendermonde area is a place where both the downstream coastal and the upstream river flow boundary conditions interact and jointly control the flood risk. Downstream of this area, the coastal level changes include both the sea level rise and storm surge changes due to climate change impacts on the wind climate over the North Atlantic and North Sea region. Upstream of the Dendermonde area lies the Dender river which introduces an extra pressure on the Dendermonde area. The impact analysis was performed using a hydrodynamic model that accounts for such changes.

From the water level impact results, it was deduced that the sea level rise and surges are by far the most important factors when evaluating the flood risk in the Dendermonde region. When the extreme changes for the three main boundary conditions (mean sea level, surge and upstream runoff) coincide, the impact is disastrous. For example, water levels at Dendermonde were simulated to change around +1.8 m for return periods in the range between 100 and 10000 years and the scenario with +0.6 m mean scenario for sea level rise, +21% high scenario change in surge levels, and +30% high scenario change in upstream flow. For the high mean sea level rise scenario of +2 m, the changes are even much higher. Water management plans, which are currently under way such as the Sigma Plan, can be informed from such studies so as to test the robustness of the proposed flood mitigation measures.

The coastal division (Afdeling Kust) of the Authorities of Flanders, Ministry for Mobility and Public Works plans to consider these correlated coastal-inland climate scenarios when implementing the Masterplan Coastal Safety (Masterplan Kustveiligheid) as approved by the Government of Flanders on 10 June 2011.

4.6 Impacts of climate change versus land use/urbanization trends

In order to separate in the river flow trends the contributions from climate change and the non-meteorological trends (i.e. land use trends), the impact of recent land use trends also has been investigated (Poelmans et al., 2011; Vansteenkiste, 2012). Poelmans et al. (2011) developed urban expansion scenarios for Flanders by extrapolation of satellite based land use maps of 1976; 1988 and 2000. It was observed that the fraction urban built up land approximately doubled from 1976 till 2000. The fraction paved land was assessed to be 4 to 5% in 1976 till about 10% in 2000. High-mean-low urban expansion scenarios were developed, the impact studied on the river flood hazard along the Molenbeek catchment (river Dijle basin) and compared with the impact of the high-mean-low CCI-HYDR climate scenarios. The results suggest that possible future climate change is the main source of uncertainty affecting changes in peak flow and flood extent. The different climate change scenarios result in a broad range of results for the peak flows: future peak flows increase (on average) with more than 30% under the wet summer climate scenario, while they decrease with almost 18% under the dry climate scenario. The urban expansion scenarios show a more consistent trend. An increase of the build-up land in the catchment with 70–200% caused an increase of the peak flows of 6–16%. The potential damage related to a flood is, however, mainly influenced by land cover changes that occur in the floodplain.

Vansteenkiste (2012) simulated the same climate and urban expansion scenarios for the catchment of the Grote Nete – Grote Laak, inter-comparing the models MIKE SHE and WetSpa. Both models simulated an increase in high flows and volumes and a reduction in baseflow rates and volumes, particularly during summer periods, as a consequence of the urban expansion scenarios. They concluded that the specific rate and magnitude of the simulated impact highly depends on the model (structure). The WetSpa model simulated an increase of the peak flows almost 4 times higher than the MIKE SHE model, which is particularly attributed to the consideration of an imperviousness factor for the urban land in the WetSpa model structure. The changes at the internal flow gauging stations, estimated with the MIKE SHE model, largely vary in function of the local characteristics. The changes were positively correlated to the percentage increase in urban land, but also to the total proportion of urban land.

4.7 *Socio-economic and ecological impacts of changes in river floods*

Next to the hydrological and hydraulic impacts, also the wider implications of the changes in flood hazard to the environment and the society were studied. This was for river floods along the Belgian rivers Dender and Ourthe done by the research project ADAPT for BelSPO (Giron et al., 2010). One of the environmental aspects that is not considered in the ADAPT project is the impact on the water quality. This aspect was taken up in the cluster project SUDEM-CLI for BelSPO, considering the Grote Nete – Grote Laak catchment (Staes et al., 2012). Also the implications to society, water managers and policy makers were assessed in that project.

Interdisciplinary cooperation was also the focus of the EU education project LECHe that developed e-learning modules on climate change and the lived experiences. Master dissertation projects were set up where students from different disciplines and with different methodological backgrounds worked together on a case to study the multi-sectorial impact of climate change on water availability at river basin level (Wilson et al., 2011; Willems et al., 2012b).

5 MORE INFORMATION

More detailed technical information about the CCI-HYDR project can be found in the five technical reports on the project website: http://www.kuleuven.be/hydr/CCI-HYDR.htm.

REFERENCES

Baguis, P., Roulin, E., Willems, P. & Ntegeka, V. 2009. Climate change scenarios for precipitation and potential evapotranspiration over central Belgium. *Theoretical and Applied Climatology*: 99(3–4), 273–286.

Baguis, P., Roulin, E., Willems, P. & Ntegeka, V. 2010. Climate change and hydrological extremes in Belgian catchments. *Hydrol. Earth Syst. Sci. Discuss.*: 7, 5033–5078.

Boukhris, O., Willems, P. & Vanneuville, W. 2008. The impact of climate change on the hydrology in highly urbanized Belgian areas. In: Water & Urban Development Paradigms: Towards an integration of engineering, design and management approaches, J. Feyen, K. Shannon, M. Neville (ed.), CRC Press, Taylor & Francis Group, 271–276.

Brouwers, J., Peeters, B., Willems, P., Deckers, P., De Maeyer, Ph., De Sutter, R. & Vanneuville W. 2009. MIRA 2009: Milieuverkenning 2030, Hoofdstuk 11 "Klimaatverandering en waterhuishouding", Milieurapport Vlaanderen, Vlaamse Milieumaatschappij, 283–304.

Bultot, F., Coppens, A., Dupriez, G.L., Gellens, D. & Meulenberghs, F. 1988. Repercussions of a CO_2 doubling on the water cycle and on the water balance – A case study for Belgium. *Journal of Hydrology*: 319–347.

Dams, J., Salvadore, E., Van Daele, T., Ntegeka, V., Willems, P. & Batelaan, O. 2012. Spatio-temporal impact of climate change on the groundwater system. *Hydrology and Earth System Sciences*: 16(5), 1517–1531.

Gellens, D. & Roulin, E. 1998. Stream flow response of Belgian Catchments to IPCC Climate change scenarios. *Journal of Hydrology*: 210, 242–258.

Giron, E., Joachain, H., Degroof, A., Hecq, W., Coninx, I., Bachus, K., Dewals, B.J., Ernst, J., Pirotton, M., Staes, J., Meire, P., De Smet, L. & De Sutter, R. 2010. Towards an integrated decision tool for adaptation measures – Case study: floods – ADAPT. Final report for Belgian Science Policy Office, 124 p.

Ntegeka, V. & Willems, P. 2008a. Trends and multidecadal oscillations in rainfall extremes, based on a more than 100 years time series of 10 minutes rainfall intensities at Uccle, Belgium. *Water Resources Research*: 44, W07402.

Ntegeka, V. & Willems, P. 2008b. CCI-HYDR Perturbation Tool: a climate change tool for generating perturbed time series for the Belgian climate. Manual version December 2008, KU Leuven – Hydraulics Section & Royal Meteorological Institute of Belgium, December 2008, 7 p.

Ntegeka, V., Decloedt, L-C., Willems, P. & Monbaliu, J. 2012. Quantifying the impact of climate change from inland, coastal and surface conditions. In: Comprehensive Flood Risk Management – Research for policy and practise (Eds. F. Klijn & T. Schweckendiek), CRC Press, Taylor & Francis Group, Leiden, The Netherlands.

Ntegeka, V., Willems, P., Roulin, E. & Baguis, P. (in revision). Developing tailored climate change scenarios for hydrological impact assessments. *Journal of Hydrology*: in revision.

Poelmans, L., Van Rompaey, A., Ntegeka, V. & Willems, P., 2011. The relative impact of climate change and urban expansion on river flows: a case study in central Belgium. *Hydrological Processes*: 25(18), 2846–2858.

Roulin, E. & Arboleda, A. 2002. Integrated modelling of the hydrological cycle in relation to global climate change – The Scheldt river Basin. Complement to the Final Report of the project GC/34/08A in the framework of the "Global change and sustainable development" program of the Belgian Federal Services of Scientific, Technical and Cultural Affairs, 45 p.

Staes, J., Willems P., Marbaix, Ph., Vrebos, D., Bal, K. & Meire, P. 2011. Impact of climate change on river hydrology and ecology: A case study for interdisciplinary policy oriented research. Final report for Belgian Science Policy Office, U. Antwerp – ECOBE, KU Leuven – Hydraulics Section & UCL – ASTR, July 2011, 103 p.

Vansteenkiste, Th. 2012. Climate change and urban expansion impact on high and low flows and the overall water availability. PhD manuscript, KU Leuven – Faculty of Engineering, December 2012.

Vansteenkiste, Th., Tavakoli, M., Ntegeka, V., Willems, P., De Smedt, F. & Batelaan, O. 2012. Climate change impact on river flows and catchment hydrology: a comparison of two spatially distributed models. *Hydrological Processes*: doi: 10.1002/hyp.9480, in press.

Vansteenkiste, Th., Tavakoli, M., Ntegeka, V., De Smedt, F., Batelaan, O., Pereira, F. & Willems, P. In revision. Intercomparison of runoff predictions by lumped and distributed models before and after climate change. *Journal of Hydrology*: in revision.

Vanuytrecht, E., Raes, D. & Willems, P. 2011. Considering sink strength to model crop production under elevated atmospheric CO_2. *Agricultural and Forest Meteorology*: 151(12), 1753–1762.

Vanuytrecht, E., Raes, D., Willems, P. & Geerts, S. 2012. Quantifying field-scale effects of elevated carbon dioxide concentration on crops. *Climate Research*: 54, 35–47.

Willems, P. 2009. Actualisatie en extrapolatie van hydrologische parameters in de nieuwe Code van Goede Praktijk voor het Ontwerp van Rioleringssystemen. KU Leuven – Afdeling Hydraulica, Eindrapport voor VMM-Afdeling Operationeel Waterbeheer, september 2009, 79 p.

Willems, P. Submitted. Revision of urban drainage design rules based on extrapolation of design rainfall statistics. Submitted.

Willems, P. & Yiou, P. 2010. Multidecadal oscillations in rainfall extremes. *Geophysical Research Abstracts*: Vol. 12, EGU2010-10270.

Willems, P. & Vrac, M. 2011. Statistical precipitation downscaling for small-scale hydrological impact investigations of climate change. *Journal of Hydrology*: 402, 193–205.

Willems, P., De Bruyn, L., Maes, D., Brouwers, J. & Peeters B. 2009. NARA 2009: Natuurverkenning 2030, Natuurrapport Vlaanderen, Hoofdstuk 2 "Klimaat". Instituut voor Natuur – en Bosonderzoek, rapport INBO.M.2009.7, 55 66.

Willems, P., Ntegeka, V., Baguis, P. & Roulin, E. 2010. Climate change impact on hydrological extremes along rivers and urban drainage systems. Final report for Belgian Science Policy Office, KU Leuven – Hydraulics Section & Royal Meteorological Institute of Belgium, December 2010, 110 p.

Willems, P., Olsson, J., Arnbjerg-Nielsen, K., Beecham, S., Pathirana, A., Bülow Gregersen, I., Madsen, H. & Nguyen, V-T-V. 2012a. Impacts of climate change on rainfall extremes and urban drainage. IWA Publishing, 252 p., Paperback Print ISBN 9781780401256; Ebook ISBN 9781780401263.

Willems, P., Kroeze, C. & Lohr, A. 2012b. The essential role of natural sciences in climate change Masters education. *International Journal of Innovation and Sustainable Development*: 6(1), 31–42.

Wilson, G., Abbott, D., De Kraker, J., Pérez, P., Terwisscha van Scheltinga, C. & Willems, P. 2011. The lived experience of climate change: creating Open Educational Resources and virtual mobility for an innovative, integrative and competence-based track at Masters level. *International Journal of Technology Enhanced Learning*: 3(2), 111–123.

Transboundary Water Management in a Changing Climate – Dewals & Fournier (Eds)
© 2013 Taylor & Francis Group, London, ISBN 978-1-138-00039-1

Impact of climate change on inundation hazard along the river Meuse

B. Dewals
Research Group Hydraulics in Environmental and Civil Engineering (HECE),
University of Liege (ULg), Belgium

G. Drogue
Centre de Recherches en Géographie (CEGUM), Université de Lorraine, Metz, France

S. Erpicum, M. Pirotton & P. Archambeau
Research Group Hydraulics in Environmental and Civil Engineering (HECE),
University of Liege (ULg), Belgium

ABSTRACT: As a part of the effort to scientifically inform the development of the adaptation strategy for the Meuse basin, we detail hereafter the generation of integrated climate and hydrological scenarios for the whole basin. We also present the setup of a first coordinated hydraulic modelling from spring to mouth of the river Meuse. The latter has enabled to compute the range of change in inundation hazard under the "wet" transnational hydrological scenario for the time slices 2021–2050 and 2071–2100. A significantly higher impact of climate change has been found in the middle part of the Meuse basin, compared to the upper and the lower parts. These conclusions have been further confirmed by a refined analysis conducted for a 100 km-long stretch of the river Meuse crossing the Belgian-Dutch border.

1 INTRODUCTION

Flooding is the most common natural hazard and third most damaging globally (Wilby and Keenan, 2012). Since flood risk is expected to further increase as a result of environmental changes, including climate change, adaptation strategies need to be developed to manage the risk affecting populations and goods.

In their review of flood mapping practices in Europe, Van Alphen et al. (2009) highlight the need for more uniform approaches in flood (risk) assessments and mapping since many European rivers are part of transboundary basins. Similarly, Becker et al. (2007) compared flood management factors in the German and Dutch parts of the Rhine basin and conclude on the need for more efficient transboundary flood management. They also suggest means to develop a common vision for future flood strategies and implement flood management issues. Van Pelt et al. (2011) showed that adequately capturing the transnational character of the river basins remains a challenging research question: whereas integrated analysis at the full river basin level, rather than within the boundaries of the riparian countries, would offer new adaptation opportunities, it will also meet many practical challenges.

Like many other basins in Europe, the Meuse basin is transnational. With a drainage surface of 35,000 km², it covers parts of France, Belgium, The Netherlands, Germany as well as a small portion of Luxembourg. The Meuse is a rain-fed river with limited groundwater storage capacity to buffer precipitations. As a result, its discharge fluctuates considerably with seasons. For instance, measured flow rates in Liege may be as low as 20 m³/s during low flows, whereas they exceeded 3000 m³/s during winter 1993.

Since 2002, the countries of the Meuse basin have been cooperating through the International Meuse Commission (IMC) to coordinate the implementation of the Water Framework Directive (2000/60/EC) and, more recently, the EU Floods Directive (2007/60/EC). In the framework of the on-going AMICE project, a basin-wide coordinated strategy is being developed to cope with hydrological impacts of climate change, including floods and low flows.

However, a lack of knowledge remains concerning the influence of climate change on flood risk in the Meuse basin, specifically on inundation hazard (including flood discharges and inundation characteristics) which constitutes a crucial input for developing a coordinated adaptation strategy for the whole basin. Among others, Leander et al. (2008) evaluated the effect of climate change on flood discharge of the Meuse by using precipitation and temperature data from three regional climate model (RCM) experiments, driven by two different global circulation models (GCM). The HBV rainfall-runoff model was used for the control climate (1961–1990) and the SRES-scenario A2 (2071–2100). It was found that the changes in the flood discharges were highly sensitive to the driving GCM.

As the adaptation strategy to be developed for the Meuse basin is intended to be truly integrated at the level of the transboundary basin, it requires a significant degree of coordination and/or harmonization of the existing regional tools and methodologies for flood risk analysis, such as climate and hydrological scenarios, hydraulic modelling and impact modelling. In particular, existing hydrological scenarios were too heterogeneous and too sporadic to be used at the basin scale. Therefore, the AMICE project has contributed to coordinating and/or harmonizing the modelling tools and methodologies throughout the basin.

In this paper, we focus on the development of integrated climate and hydrological scenarios, as well as on the setup of coordinated hydraulic modelling for the whole Meuse and some tributaries.

2 METHODOLOGY

Evaluating the impact of climate change on inundation hazard requires that climate projections are downscaled to the relevant scales characterizing hydrological processes and inundation flows in the floodplains (Table 1). Therefore, this study involves three main steps, namely the development of integrated climate and hydrological scenarios (steps 1 and 2), as well as hydraulic modelling along river Meuse and some tributaries (step 3).

2.1 *Integrated climate scenarios*

Following a review of climate experiments used within the Meuse basin, showing the lack of bias corrected climate simulations at the Meuse basin scale, the *delta change* approach was first applied to existing national climate scenarios. Next, common transnational climate scenarios with high resolution time series were derived (Drogue et al., 2010).

The greenhouse gases (GHG) emission scenarios commonly used in climate change studies have been developed by the Intergovernmental Panel on Climate Change (IPCC) since 1996 and they have been described in the Special Report on Emission Scenarios (SRES). For each group of scenarios, one scenario has been selected as a reference (A1B, A2, B1 and B2). These are the most widely used scenarios for GCM simulations and for impact studies of climate change.

Based on GCM /RCM simulations forced with IPCC SRES emission scenarios, each national meteorological institute has provided seasonal trends for the future climate (Δ in % for rainfall change and in °C for air temperature variation in winter, spring, summer and autumn).

Although many existing climate change studies provide insights for the end of the century, decision makers also need information on the short and medium terms. Therefore, two time horizons were considered: 2021–2050 and 2071–2100. The 30 years span was used because climatological data are generally available for a 30 years long reference period.

The climate of these two periods has been compared to a reference period (1971–2000). Monthly or even daily data series would have provided an even better insight, but these were not yet validated

Table 1. Typical space and time scales in climate, hydrological and hydraulic modelling.

	Space scale	Time scale
Global Circulation Models	$\sim10^5$ m	$\sim10^5$ s
Regional Climate Models	$\sim10^4$ m	$\sim10^5$ s
Rainfall-runoff models (catchment-scale)	10^2–10^3 m	10^3–10^4 s
Hydraulic and impact models (river- and floodplain-scale)	1–10 m	10^2–10^3 s

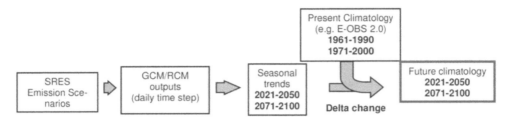

Figure 1. The delta change approach as carried out in the Amice project.

when the work started. The delta change approach has been used to modify the climate parameters on the reference period through seasonal perturbation factors (Figure 1).

All global circulation models agree that air temperature will increase in Europe in the coming decades. In contrast, rainfalls are either expected to increase or to decrease in the Meuse area, depending on the models. To account for such uncertainties, the delta change approach has been applied to create one "wet" and one "dry" climate scenarios for each period and for each national sub-basin. In between these two scenarios lies a wide range of possible futures.

Since the seasonal trends obtained for each national sub-basin presented significant heterogeneities, a transnational scenario was developed by weighting national trends according to the drainage area of each sub-basin. This enables to maintain downstream consistency of discharges, especially at national and regional borders.

2.2 *Integrated hydrological scenarios*

For nine selected gauging stations, rainfall-runoff modelling was conducted to estimate the evolution of flood and low-flow discharges during the 21st century. No single rainfall-runoff model is available to cover the whole Meuse basin with a sufficient level of details. The different rainfall-runoff models used in each sub-basin are given in Table 2. These models compute river discharges using basically the same input climate data (air temperatures, potential evapotranspiration and precipitation) and some characteristics of the basin and river (slope, land cover, etc.).

Besides a control run (1971–2000 or 1961–1990), the two time horizons 2021–2050 and 2071–2100 were simulated. The considered gauging stations are located in four countries:

- four stations along the French part of the river Meuse: Saint-Mihiel, Stenay, Montcy-Notre-Dame and Chooz;
- one station at the Belgian-Dutch border: Sint Pieter;
- four stations on the Belgian and German right-side tributaries: Gendron (river Lesse), Chaudfontaine (river Vesdre), Stah (river Rur) and Goch (Niers triver).

Table 2. Rainfall-runoff models used in the different sub-basins of river Meuse.

Sub-basin	Model
France	AGYR and GR4J
Belgium (Wallonia)	EPIC-Grid
Belgium (Flanders)	TOPModel and MIKE11 Maas
The Netherlands	HBV
Germany	NASIM and GR4J

In order to define the values of discharges for different return periods, a statistical distribution has been fitted to the observed and simulated discharge series. The sampling method of annual maximum discharges was used, as it is the most common method used to evaluate quantiles of flood discharges. In most cases, the parameters of the statistical distributions were estimated through the maximum-likelihood method. Different theoretical statistical distributions (Gumbel, Weibull, etc.) were used for calculating the quantiles of annual winter hourly maximum discharge value (e.g., Qhx_{100}). For each gauging station, the theoretical statistical distribution was selected according to the fitting quality between observed and calculated quantiles.

2.3 Coordinated hydraulic modelling

Coordinated hydraulic modelling was first performed from spring to mouth of the river Meuse, based on exchanges of boundary conditions between the existing models in the different regions. Next, two very similar 2D unsteady models were applied to conduct a more refined analysis of a selected stretch of about 100 km crossing the Belgian-Dutch border.

2.3.1 Coordinated modelling from spring to mouth of river Meuse

Hydraulic models are available in each region of the Meuse basin. They are either commercial ones or academic codes. However, significant differences between those models have been identified, reflecting differences in the characteristics of the basin, including:

- in terms of spatial representation, the model range from fully one-dimensional, based on cross-sections even in the floodplains (e.g., in France), up to fully two-dimensional description based on laser altimetry and sonar bathymetry (e.g., in Wallonia);
- time description also differs between the models (either unsteady or run in steady mode).

Based on a detailed review of existing modelling procedures, a coordinated methodology for hydraulic modelling has been developed (Detrembleur et al., 2012). The methodology involves the following key aspects:

- consistency of bathymetry has been ensured across the borders, including latest dredging works; although the data in the different regions were not collected at the same period.
- regional hydrological time series and statistical analysis have been compared to derive consistent discharge values for a wide range of return periods.
- a coordinated procedure has been elaborated for running the hydraulic models.

Following this newly developed procedure, the hydraulic models from the different regions have been run in parallel but not coupled online. Nonetheless, consistent hydraulic results across the borders have been obtained in just two runs:

- in the first one, necessary boundary conditions have been deduced from extrapolation of measured stage-discharge relationships;

- in the second one, the boundary conditions have been refined if necessary, using the results of the first run of the neighbouring models.

This coordinated modelling methodology has enabled to conduct the first hydraulic simulation from spring to mouth of river Meuse. The modelling has been run for the 100-year flood in the base scenario (present situation), as well as for the "wet" hydrological scenario (section 2.2). The two aforementioned time horizons are also considered here: 2021–2050 and 2071–2100.

2.3.2 *Refined analysis*

The characteristics of the hydraulic models used in the different regions differ significantly. In particular, the model used in the Walloon part of the Meuse basin was fully two-dimensional, but was run in steady mode. In contrast, the model used in the Dutch part of the Meuse was one-dimensional and unsteady.

Therefore, a refined analysis has been undertaken along a 100 km long transnational section of the Meuse, between Ampsin and Maaseik, using very similar models for the Walloon and the Dutch parts: WOLF 2D (Belgian part) and WAQUA (Dutch part), which are both 2D and run in unsteady mode. This has enabled to quantify the damping of the flood waves and to assess the relevance of using a steady model in this part of the Meuse basin.

Additionally to the peak discharge, the shape of the whole flood wave is necessary to prescribe the upstream boundary conditions of unsteady hydraulic simulations. The different methodologies to generate synthetic flood waves available in Wallonia, Flanders and the Netherlands were compared and tested to generate a 100-year synthetic flood wave at the Belgian-Dutch border. The Walloon methodology was selected due to the high degree of similarity between the obtained flood wave and the measured hydrograph during the 1993 flood, its return period being the closest to 100 years among available records. Using the Walloon methodology is also of particular relevance, since the selected methodology is dedicated to be applied to generate inflow flood waves mainly in Ampsin and for river Ourthe in Liege.

Consistently with the integrated hydrological scenarios, the whole flood waves were assumed to be increased by, respectively, 15% and 30% for the time horizons 2021–2050 and 2071–2100.

3   RESULTS AND DISCUSSION

3.1   *Integrated climate scenarios*

Under the wet transnational scenario (Figure 2), air temperatures are expected to increase by 1.3°C to 2.9°C, with little difference between the seasons. Increases in air temperatures are higher for the dry scenario, especially in summer. According to the wet scenario, precipitations are expected to increase in winter (+11% in 2021–2050 and +25% in 2071–2100) whereas, even under this scenario, precipitations are expected to decrease in summer. Under the dry scenario, precipitations tend mostly to decrease, or to remain unchanged.

In order to check the consistency of the approach, the results for the transnational scenarios were compared to the RCM simulations produced by the European FP5 PRUDENCE project (De Wit et al., 2007). Both results match relatively closely, which reinforces the methodology used.

3.2   *Integrated hydrological scenarios*

To analyse the changes in flood discharges, the annual winter hourly maximum discharge, noted Qhx was selected as representative hydrological variable. The considered return periods are 2, 5, 10, 25, 50 and 100, as well as 250 and 1250 for the lower part of the basin. The results presented here focus on the 100-year return period (Qhx_{100}).

In order to characterize the evolution of the flood discharges, a perturbation factor was calculated at the nine aforementioned gaugingstations for both considered time horizons (2021–2050 and 2071–2100) and for the wet, dry as well as national climate scenarios. The perturbation factor is

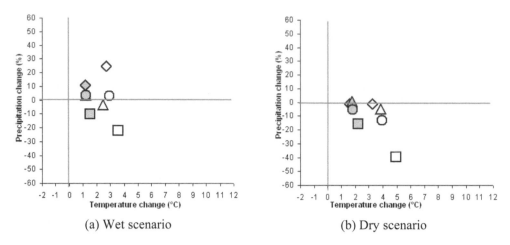

(a) Wet scenario   (b) Dry scenario

Figure 2.   Seasonal trends in precipitation (%) and air temperature (°C) for the transnational scenarios and for the two time slices (grey: 2021–2050; white: 2071–2100; ◇ = winter, Δ = spring, □ = summer, ○ = autumn).

Table 3.   Perturbation factors obtained for the hourly winter centennial flood peak (Qhx_{100}), wet and dry climate scenarios, at the gauging stations located along the main course of river Meuse.

Time horizon	Scenario	St-Mihiel	Stenay	Montcy	Chooz	St-Pieter
2021–2050	Wet	1.12	1.12	1.12	1.12	1.14
	Dry	0.96	0.96	0.96	0.96	0.95
2071–2100	Wet	1.27	1.27	1.27	1.27	1.33
	Dry	0.89	0.89	0.89	0.89	0.91

defined as the ratio between the simulated values of the discharge for a given scenario and those simulated for the present climate. A value above 1 corresponds to an increase of the flood discharge value and vice-versa. Table 2 shows the perturbation factors obtained according to the transnational scenario for Qhx_{100} in the selected gauging stations located on the main course of river Meuse.

For the transnational climate scenario, the sign of the change in maximum discharge is logically homogeneous across the basin: an increase (decrease) in discharge is expected for the wet (dry) scenario. The range of change is more important for the end of the century.

Based on the results of Table 3 and discussions with a panel of experts and stakeholders in the Meuse basin, hydrological scenarios for the whole Meuse basin were defined (Drogue et al., 2010). The most extreme hydrological scenario considered for flood risk analysis was derived from the transnational wet climate scenario. In agreement with Table 3, it assumes an increase in Qhx_{100} values of +15% for 2021–2050 and +30% for 2071–2100.

3.3 *Changes in inundation hazard*

The results of hydraulic modelling results have been analysed at two complementary scales, namely the reach scale and the local scale (hotspots).

At the reach scale, the hydrological impacts of climate change have been evaluated in terms of their effects on hydrograph damping and peak discharges, water levels, extent of inundated areas and volume stored in the floodplains (Detrembleur et al., 2012). In particular, the integrated hydraulic modelling conducted has revealed a strong spatial pattern in the sensitivity of river stages

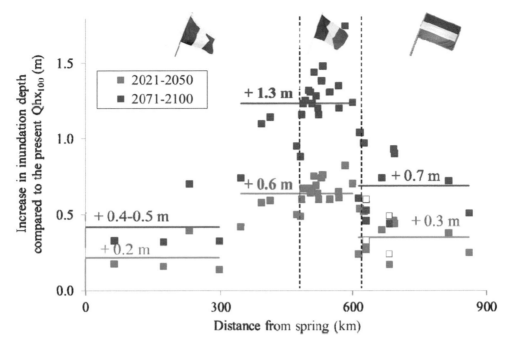

Figure 3.   Change in flood levels compared to the present 100-year flood.

with respect to changes in flood discharge (Figure 3): the influence of a similar change in flood discharge is found to be approximately twice stronger in the central part of the basin (between Sedan and Monsin) compared to the upper and lower parts of the basin, respectively upstream of Sedan and downstream of Monsin. This finding can be easily explained by considering the main characteristics of the Meuse basin: both the upper and the lower parts of the basin (including lowlands in The Netherlands) are characterised by relatively wide floodplains with large storage capacity; whereas, in the central part of the basin (Ardennes massif) the river valleys are steeper and narrower, leading to limited storage capacity in the floodplains. As a result, river stages are indeed expected to show a higher sensitivity in the central part of the basin. We also checked that this abrupt change in the sensitivity of water stages does not coincide with a change in the hydraulic models used. As it was not the case, we come to the conclusion that this finding is not affected by such kind of numerical artefact.

Analysis of a hotspot in France (Charleville-Mézières) highlights that, although extra flooded areas due to climate change have been found generally limited in their extent, they may nevertheless lead to serious impacts when key assets are situated within the future flood-prone area (e.g., a town hall in which the crisis management headquarter is located). As shown in Figure 4, the hotspots in Wallonia reveal that existing flood defences of the main cities along river Meuse (Namur and Liege) may not be considered as "climate-proof". Indeed, although these cities are basically protected for present climate conditions, they would in contrast be exposed to huge flooding as a result of climate change. The common Flemish and Dutch hotspot (stretch of river Meuse along the border) emphasizes the importance of representing accurately protection structures (e.g., dikes) in the hydraulic modelling.

The refined unsteady 2D hydraulic simulations have shown that only a very limited damping of the flood waves is obtained along the simulated reaches (Dewals et al., 2013):

• for the present Qhx_{100}, a maximum of $\sim$1% damping of the peak discharge in the Walloon part and another $\sim$1% damping for the Dutch part;

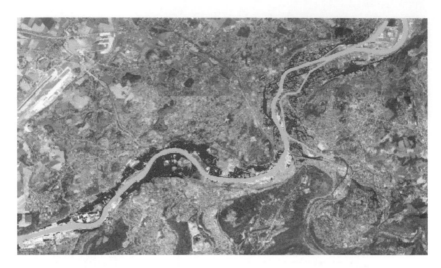

Figure 4.   Inundation extent in the area of Liege for a 100-year flood in the present climate (■), additional extent for the 2021–2050 period (wet scenario) (▪), additional extent for the 2071–2100 period (wet scenario) (■). Hydraulic computations were performed with WOLF 2D in unsteady mode.

- for $Qhx_{100} + 15\%$, the damping of the peak discharge remains around 1% in the Walloon part, but reaches about 3% in the Dutch part;
- for $Qhx_{100} + 30\%$ the damping of the peak discharge is of maximum 3 to 4% both in Wallonia and in the Netherlands.

From the perspective of the variation in peak discharge, the study enables to conclude that using the Walloon model WOLF 2D in steady mode makes perfectly sense. Nonetheless, extensive overestimations by the steady model of the inundated extent and of the stored volume in the floodplains have been highlighted.

4   CONCLUSION

Integrated climate change scenarios for the Meuse river basin have been derived from a weighted-average of national and regional climate scenarios obtained by the delta change method. To take into account the uncertainties on future climate, a "wet" and a "dry" scenario have been defined for the time horizons 2021–2050 and 2071–2100. They were used to force different rainfall-runoff models, from which integrated hydrological scenarios were produced. Under the wet transnational scenario, flood discharges are expected to increase by, respectively, 15% (2021–2050) and 30% (2071–2100).

Coordinated hydraulic modelling was conducted from spring to mouth of river Meuse. It has revealed a strong spatial pattern in the sensitivity of flood levels with respect to changes in flood discharge: the increase in flood levels is about twice higher in the central part of the basin compared to the upper and lower parts. The topographic characteristics of the valley explain this change, which does not result from a change in the model used. The hydraulic modelling has also provided the changes in inundation extent and volume stored in the floodplains.

To investigate the sensitivity of the results with respect to modelling assumptions differing from one region to the other, a refined analysis has been performed for a 100 km long section of river Meuse crossing the Belgian-Dutch border. Based on detailed 2D unsteady simulations, this analysis has confirmed the marginal damping of flood waves in this section of the Meuse and, consequently, the relevance of using steady simulations for inundation modelling in this area.

The results of this research also emphasize the need for more analysis concerning several key aspects. Only few SRES scenarios were considered and the integrated scenarios do not reflect the whole uncertainty range, especially a low probability/high impact climate scenario was not considered. More hydrological and hydraulic modelling will also be needed, such as to investigate the influence of landuse change in the catchment and in the floodplains, as well as to evaluate various adaptation measures (e.g., adapted spatial planning). The possible damping of flood waves in reaches upstream of Ampsin should be studied as well.

The 2D unsteady model set up so far constitutes a tool of primary interest, which is readily available to design and evaluate protection measures for future flood-prone areas such as Liege. It also constitutes a very valuable input for impact assessment and risk modelling (e.g., Ernst et al., 2010), as well as for the elaboration of the Meuse adaptation strategy.

ACKNOWLEDGEMENTS

This research was carried out in the framework of the Amice project funded under the NWE Interreg IVB Programme. The authors from ULg also gratefully acknowledge the "Service Public de Wallonie" (SPW) for the Digital Surface Model and the hydrological data.

REFERENCES

Becker, G., Aerts, J. & Huitema, D. 2007. Transboundary flood management in the Rhine basin: Challenges for improved cooperation. *Water Science and Technology*, 56, 125–135.

De Wit, M. J. M., Van den Hurk, B., Warmerdam, P. M. M., Torfs, P. J. J. F., Roulin, E. & Van Deursen, W. P. A. 2007. Impact of climate change on low-flows in the river Meuse. *Climatic Change*, 82, 351–372.

Detrembleur, S., Dewals, B., Fournier, M., Becker, B., Guilmin, E., Moeskops, S., Kufeld, M., Archambeau, P., De Keizer, O., Pontegnie, D., Huber, N. P., Vanneuville, W., Buiteveld, H., Schüttrumpf, H. & Pirotton, M. 2012. Effects of climate change on river Meuse: hydraulic modelling from spring to mouth. Scientific report of the AMICE Project (WP1-Action 6).

Dewals, B., Archambeau, P., Huismans, Y., De Keizer, O., Detrembleur, S., Buiteveld, H. & Pirotton, M. 2013. Effects of climate change on river Meuse: hydraulic modelling from Ampsin to Maaseik. Scientific report of the AMICE Project (WP1-Action 6).

Drogue, G., Fournier, M., Bauwens, A., H., B., Commeaux, F., Degré, A., De Keizer, O., Detrembleur, S., Dewals, B., François, D., Guilmin, E., Hausmann, B., Hissel, F., Huber, N., Lebaut, S., Losson, B., Kufeld, M., Nacken, H., Pirotton, M., Pontégnie, D., Sohier, C. & Vanneuville, W. 2010. Analysis of climate change, high-flows and low-flows scenarios on the Meuse basin. Scientific report of the AMICE Project (WP1-Action 3).

Ernst, J., Dewals, B. J., Detrembleur, S., Archambeau, P., Erpicum, S. & Pirotton, M. 2010. Micro-scale flood risk analysis based on detailed 2D hydraulic modelling and high resolution geographic data. *Natural Hazards*, 55, 181–209.

Leander, R., Buishand, T. A., Van den Hurk, B. J. J. M. & De Wit, M. J. M. 2008. Estimated changes in flood quantiles of the river Meuse from resampling of regional climate model output. *Journal of Hydrology*, 351, 331–343.

Van Alphen, J., Martini, F., Loat, R., Slomp, R. & Passchier, R. 2009. Flood risk mapping in Europe, experiences and best practices. *Journal of Flood Risk Management*, 2, 285–292.

Van Pelt, S. C. & Swart, R. J. 2011. Climate Change Risk Management in Transnational River Basins: The Rhine. *Water Resources Management*, 25, 3837–3861.

Wilby, R. L. & Keenan, R. 2012. Adapting to flood risk under climate change. *Progress in Physical Geography*, 36, 348–378.

Transboundary Water Management in a Changing Climate – Dewals & Fournier (Eds)
© 2013 Taylor & Francis Group, London, ISBN 978-1-138-00039-1

Impacts of future floods and low flows on the economy in the Meuse basin

B. Sinaba, R. Döring, M. Kufeld & H. Schüttrumpf
Institute of Hydraulic Engineering and Water Resources Management RWTH Aachen

A. Bauwens
Gembloux Agro-Bio Tech (ULG)

ABSTRACT: Climate change in Western Europe is projected to result in more humid winters and drier summers. Further, the severities of floods and low flows are assumed to increase in the future. The impacts of these events could lead to adverse consequences on the economy. Given this framework, within the AMICE project, the impacts of future floods and low flows will be analyzed. A flood risk analysis in the Meuse basin is conducted taking into account future climate scenarios. Further, the impacts of future droughts and low flows are analyzed for the economic sectors energy production, agriculture and navigation.

1 INTRODUCTION

The issues of flood risk have been attracting attention in recent years and have moved up on the political and scientific agendas following increased frequency and severity of flood events. Additionally, more frequent periods of low flows in dry summer months have been observed, culminating in the 2003 dry season in Western Europe. As climate change in Western Europe is predicted to result in more humid winters and drier summers, which both are expected to be accompanied by a higher frequency of extremes, the severities of floods and low-flows are assumed to increase in the future. Due to limited natural storage capacity in the Meuse basin a direct link exists between climate evolutions and changes in high and low-flows, putting at risk the assets of the basin. Given this framework the provided study deals with methodologies to quantify the impacts of floods and low flows.

In the AMICE project future hydrological scenarios (FS) on floods (FS_{wet}) and low flows (FS_{dry}) are estimated for the time horizons 2021–2050 (FSI) and 2071–2100 (FSII). For these dry and wet future scenarios FS I and FS II, the economic consequences are examined and compared with the impacts of the hydrological scenarios of the present state PS, representing the time slice between 1971 and 2000. The future hydrological conditions of floods and low flows are presented in Drogue (2010). Due to the underlying different hydrologic and cause-effect perspectives, the impacts due to low flows and floods are studied separately. In section 2 methodologies are introduced to quantify the impacts of the future hydrological conditions. The results are presented in section 3. In consideration of the uncertainties in the overall chain of the flood and as well for the low flow impact assessment, it is avoided to present the results in absolute monetary terms but rather in percentages in all considered sectors.

2 METHODOLOGY

2.1 Impacts of future floods

The impacts of future floods are examined via flood risk analysis. In general, the risk is defined as the product of the probability of occurrence of an undesired event and the magnitude of its

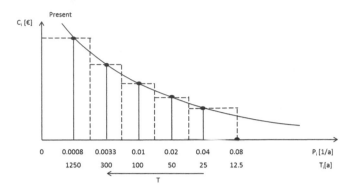

Figure 1.   Risk curve associated to return periods representing the present state.

consequences (Hall et al., 2004). The undesired event is considered as the flood event and its probability of recurrence (or flood frequency) is described by the associated return period. The consequences accompanied by the flood depend on the vulnerabilities and are expressed in terms of economic damage assigned to this event. The combined information are then plotted as an exceedance probability – loss curve which is a conventional method to illustrate flood risk. According to Kaplan and Garrick (1980) this curve is called "risk curve". Discrete values of flood damage C_i [€] are calculated for considered return periods T_i [year] and its associated flood frequency P_i [1/year]. In Figure 1 a risk curve is depicted representing the return periods and the associated flood damages of the present state PS_{wet}.

Mathematically, the integrated flood risk R is then approximated by the area under the risk curve. The approximation of the surface integral is conducted according to Bachmann (2012):

$$R = \sum_{i=1}^{k} C_i \cdot \left(\frac{P_{i+1} - P_i}{2} + \frac{P_i - P_{i-1}}{2} \right) = \sum_{i=1}^{k} C_i \cdot \left(\frac{P_{i+1} - P_{i-1}}{2} \right) \tag{1}$$

where R = integrated flood risk [€/year] ; P_i = Probability of a discrete undesirable event i [1/year]; and C_i = Consequence/Flood Damage potential of a undesirable event i [€].

According to the flood risk definition of formula (1), the risk calculation methodology is composed of an analysis of the hydraulic system and an analysis of the economic flood damage for the considered flood events. One of the inputs for the flood risk methodology is a map displaying inundated areas and flow depths in the floodplain related to a flood event and its specific return period T [year]. The hydraulic modeling and the generation of inundation maps in AMICE is described by Detrembleur, (2010). The flood damage estimation methodology is based on land use data, damage functions and specific asset values. Land use information is aggregated into 5 damage categories settlement, industry, infrastructure, agriculture and forestry. The assessment of the damage potential is then done by a superposition of the hydraulic information (inundation depth and area) and the land use data. A transformation of the inundation depth by specific damage functions which are associated with the corresponding damage category, results in the relative damage. The monetization is then realized by multiplying the relative damage by the corresponding monetary asset value. The result is the economic loss due to inundation depths in [€/m²]. The methodology of the flood damage analysis is described in Sinaba et al. (2011). Based on the damage results and the associated flood frequencies, flood risk is calculated according to formula 1 for the present state (PS_{wet}), FS I$_{wet}$ (2021–2050) and FS II$_{wet}$ (2071–2100) taking into account the future hydrological conditions. In avoidance of absolute monetary terms the risk estimates of the scenarios FS I$_{wet}$ and FS II$_{wet}$ are related to the risk estimates of the present state PS resulting in the risk increase RI_{FSI} and RI_{FSII}.

Table 1. Reduction in electricity production in [%] (Foerster & Lillistam, 2010).

Stream flow Reduction [%]	water temperature increase					
	0 K	1 K	2 K	3 K	4 K	5 K
0	0.8	1.6	3.0	5.2	8.1	11.8
10	0.8	1.7	3.1	5.2	8.2	11.9
20	0.9	1.8	3.2	5.3	8.3	12.0
30	1.4	2.2	3.7	5.8	8.7	12.4
50	6.1	6.9	8.2	10.1	12.8	16.2

2.2 Impacts of future low flows

Whereas the approaches to calculate the impacts of floods, in terms of flood risk, are sophisticated and well established, approaches to determine the impacts of droughts and low flows do exist less frequent. In the present study, the impacts of possible future drought and low flow conditions due to climate change on the economic sectors energy, agriculture and navigation are examined. The hydrological dry future scenarios $FS_{i,dry}$ applied in AMICE are characterized by a decrease in precipitation and river discharge and otherwise with an increase in air- and water temperature. These effects could lead to adverse impacts on the considered economic sectors. Our aim is to propose methodologies in order to quantify adverse implications of drought and low-flow conditions on these sectors.

2.2.1 Energy

The energy sector refers to electricity production in thermal power plants and in hydropower plants. Due to the difference in process technologies, impacts of climate change on both types of power plants are studied separately.

Thermal power plants require large water amounts for cooling purposes. During low flow periods thermal power plants are forced to operate with reduced capacity or at worst case if temperature thresholds are exceeded, the power plant has to be shut down temporarily.

The methodology applied on the thermal power plants using Meuse water, is mainly based on a study of Foerster & Lilliestam (2010). This study quantifies the reduction in electricity production via modeling of energy turnover and heat balance under changing mean annual air temperatures and thus water temperatures and river discharge. The results of this study are analyzed and adapted to the climate change projections of the AMICE future dry scenarios resulting in the correlation between discharge reduction, temperature increase and energy reduction production as shown in Table 1.

Energy production in hydropower plants is determined by the discharge and the drop height. According to Strobl & Zunic (2006), the attainable output P [kW] of the turbine can be assessed with formula

$$P = 8 \cdot Q \cdot H_n \tag{2}$$

with P = capacity of the turbine [kW], Q = discharge [m³/s] and H_n net drop height [m].

In respect of equation (2) it is obvious that a variation of the future discharge will directly affect the attainable output capacity of the hydro power plants. In consequence of decreased discharge during low flow events, hydropower plants operate below full capacity.

For the gauges close to the thermal- and hydro power plants located along the Meuse, the mean annual discharge reduction and the water temperature increase due to the future dry scenarios as indicated in Drouge (2010) are calculated. For thermal power plants, these values served then as input parameters in table 1 to interpolate the mean annual energy reduction production. Further, the reduction in electricity production in hydro power plants is the difference between the capacity resulting from formula (2), on the basis of the annual mean discharge of the present state PS_{dry} and the capacity calculated under future scenarios FS_{dry}. Hereby, the drop height H_n is assumed to be constant.

2.2.2 *Agriculture*

Agriculture is also one of the economic sectors impacted by future climate extremes. It is commonly predicted that global warming will have effects on crop yields. These effects could be positive as well as negative in accordance with the range of predicted changes and the adaptation capacity of agricultural systems.

A modeling of the main crop yields on the Meuse basin was carried out. The model used to realize the simulations is an adaptation of the EPIC (**E**rosion **P**roductivity **I**mpact **C**alculator) model EPIC-Grid. The physically-based model is able to simulate water soil plant continuum, crop growth and their uptakes and water movements in the soil. For each country in the Meuse basin, yields for the three main crops (Maize, Wheat and Barley) of the catchment are calculated for the present state (PS) and for the future dry scenarios FS_{dry}. The input data required to run the EPIC-Grid model includes (1) daily weather information, (2) soil characterization data, (3) a set of parameters characterizing the crops being grown; (4) and crop management information such as emerged plant population, air CO_2 concentration, row spacing, seeding depth and date, harvest date and fertilizer schedules (Bauwens et al., 2011).

2.2.3 *Navigation*

The Meuse is navigable over a substantial part of its total length. The flow regime in the Meuse is strongly influenced by weirs, canalization and lateral withdrawals. Weirs regulation permits to guarantee minimum water levels most of the time. And in general, navigation will not encounter problems due to insufficient water depth. Ships pass weirs via navigation locks. Problems for navigation then start to occur when the water loss due to the locking process is such that water levels and discharges cannot be guaranteed anymore.

Measures can be implemented to reduce water losses during locking process, causing extra costs. These measures consists in reducing the number of locking cycles per day and diminishing the water loss in a lock cycle by pumping (AVV, 2002). The extra costs due to future low flows are assessed for three Dutch lock complexes at Born, Maasbracht and Heel on the basis of simulations in AVV (2002). In AVV (2002), scenarios with a normal locking process are compared to scenarios with different locking strategies using the software package SIVAK. SIVAK simulates the total time and costs for each individual ship that passes the lock, and computes water losses. The simulations required the following input parameters: the number and size of the lock chambers, the water level difference between upstream and downstream of the lock, shipping intensity for the lock waiting and sailing costs per ship class.

3 RESULTS

3.1 *Impact of floods*

The increase in flood risk due to the future scenarios is depicted in Table 2.

In order to get a more detailed view of flood risk, the normalized flood risk is calculated for several reaches. Sensitive areas under future scenarios are thus identifiable. Flood risk per reach is estimated for the present state PS_{wet}, $FS\ I_{wet}$ and $FS\ II_{wet}$. These risk estimates are then related to the present state (PS) flood risk of the whole basin. Results are shown in Figure 2. Even in the upstream and the middle part of the Meuse, several reaches show a noticeable contribution to the flood risk, such as the Sedan-Aiglemont reach in France, which includes the city of Charleville-Mézières. In the Walloon region the reach between Andenne and Ampsin is found as the most sensitive. This reach includes the city Huy and several industrial areas. Although, the reach Sedan – Aiglemont is very significant, an increase in the contribution to the total future flood risk from upstream to

Table 2. Flood risk increase in [%] of the future scenarios related to the present state PS.

	FS I$_{wet}$ (2021–2050)	FS II$_{wet}$ (2071–2100)
Flood risk increase [%]	150	390

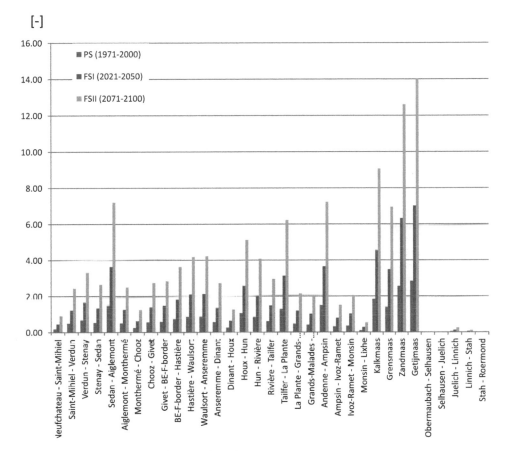

Figure 2. Normalized flood risk for several reaches.

downstream can be observed. Thus, the reaches in the lowlands of the Netherlands and along the Dutch Flemish border are showing the largest contribution to the total flood risk increase.

3.2 Impacts of low flows

3.2.1 Energy

The reduction in energy production in thermal power plants is calculated to approximately 2% for the thermal power plants for the dry future scenario I FS I$_{dry}$ as Table 3 shows. This trend worsens for the future time period from 2071–2100 as the consequence of further discharges decreases and temperature increases.

The reduction in energy production in hydropower plants is depicted in Table 4. For all hydro power stations, except Grands-Malades and Hun, the approximated reduction in energy production for future scenario FS I$_{dry}$ is more than 35%. For the future scenario FS II$_{dry}$ for these hydro power

Table 3. Reduction in electricity production in [%] (Thermal power plants).

| Power plant | annual reduction in energy production [%] | |
	FS I_{dry} (2021–2050)	FS II_{dry} (2071–2100)
Chooze	1.9	8.2
Tihang	2.0	7.5
Clauscentrale	2.0	5.8
Buggenum	3.0	8.6
Amercentrale	2.0	5.8
Dongecentrale	2.0	5.8

Table 4. Reduction in electricity production in [%] (Hydro power plant).

| Power plant | annual reduction in energy production [%] | |
	FS I_{dry} (2021–2050)	FS II_{dry} (2071–2100)
Lixhe	43.0	64.0
Monsin	57.0	73.0
Ivoz-Ramez	39.0	62.0
Ampsin-Neuville	35.0	60.0
Andenne	45.0	66.0
Grands-Malades	0	36.0
Hun	0	11.0

stations the reduction is higher than 60% compared to the present state PS. For the hydro power stations of Grands-Malades and Hun is no energy reduction predicted according to FS I_{dry}, the discharge value is still higher than the possible maximum discharge value of the turbines.

3.2.2 *Agriculture*
The variation of the future yields compared to the yield related to the present state PS is shown in Table 5. It is observed that maize crop is negatively affected by climate change in the context of the dry future scenarios despite the CO_2 fertilization effect. The decrease in maize yield is of approximately 3–4% for FS II_{dry} (only France sees a small yield increase of 2%). For the future scenario FS II_{dry}, the decrease reaches 20% and Flanders will suffer the greatest losses with 30%.

Wheat and barley have similar behaviors and are positively affected by climate change and CO_2 fertilization effect. The increase in wheat yield is from 8–15% for the future scenario FS I_{dry} and reaches 16–28% for the future scenario FS II_{dry}. The increase in barley yield is from 6 to 20% for the future scenario FS I_{dry} and from 9 to 21% for the future scenario FS II_{dry}. However, in all cases, yield variability will increase.

3.2.3 *Navigation*
The water savings of the present state, which is assumed as an average year, are related to the extra costs occurring in a dry year, representative for the future scenario FS I_{dry} and a very dry year representing the conditions of the future scenario FS II_{dry}. A description of the calculation methodology is given exemplary for the lock in Maasbracht in Sinaba et al. (2011). The conducted calculation procedure indicates that the extra costs increase with 36% for FS I_{dry} and by 1520% for FS II_{dry} as depicted in Table 6.

Table 5. Evelution of yield in [%] related to the present state (1971–2000).

Region	FS I$_{dry}$ (2021–2050)			FS II$_{dry}$ (2071–2100)		
	Maize	Wheat	Barley	Maize	Wheat	Barley
France	+2.0	+16.4	+20.3	−17.5	+27.8	+21.4
Wallonia	−3.0	+8.4	+18.5	−18.9	+17.5	+18.0
Flanders	−18.0	+7.9	+16.4	−29.3	+16.1	+18.0
Netherlands	−3.4	+12.7	+12.2	−18.8	+23.3	+11.0
Germany	−5.4	+8.4	+6.7	−21.8	+17.8	+8.8

Table 6. Increase of extra costs due to water savings in [%].

	FS I (2021-2050)	FS II (2071–2100)
Increase [%]	36.0	1520

4 CONCLUSIONS

The economic impacts of future floods and low flows have been assessed. The results have shown that an increase of flood risk due to future climate is expected. Considering the whole Meuse basin, the flood risk increase under future scenario FS I$_{wet}$ (2021–2050) is 150% and 390% under future scenario FS II$_{wet}$ (2071–2100).

While the economic impacts of floods are well assessable by means of sophisticated hydraulic models and economic damage approaches, the impacts of droughts and low flows are less studied in European countries. This document has provided a brief overview of drought and low flow impacts on the energy-, agricultural- and navigation sectors in the Meuse basin. It is shown that a decrease in discharge and a water temperature increase will affect the electricity production in thermal- and hydro power plants. Agriculture is also strongly affected by climate change, but impacts can be positive or negative depending on the considered crop. In the navigation sector, a decrease in discharge can cause economic losses due to water saving lock strategies applied during low flow periods.

Solutions will have to be studied and implemented at the local level as well as at international level. In this context of the changing climate, it can be concluded, that there is a need for an improvement to assess the impacts of drought and low flows.

REFERENCES

AVV (2002): Scheepvaartaspecten Laagwaterbeleid Julianakanaal en Lateraalkanaal. Rijkswaterstaat Adviesdienst Verkeer en Vervoer (in Dutch).

Bachmann, D. (2012): Beitrag zur Entwicklung eines Entscheidungsunterstützungssystems zur Be-wertung und Planung von Hochwasserschutzmaßnahmen. Dissertation. Aachen: Institut für Wasserbau und Wasserwirtschaft, RWTH Aachen.

Bauwens, A., Sohier, C. & Degré, A., (2011): Hydrological response to climate change in the Lesse and the Vesdre catchments: contribution of a physically based model (Wallonia, Belgium). Hydrology and Earth System Sciences, 15, pp. 1745–1756.

Cuculeanu, V., Marica, A. & Simota, C., (1999): Climate change impact on agricultural crops and adaptation options in Romania. Climate Research, 12, pp. 153–160.

Detrembleur, et al. (2012): Effect of climate change on river Meuse, Hydraulic modelling from spring to mouth. AMICE – WP1 – Action 6 – technical report, (unpublished).

Drogue, et al. (2010): Analysis of climate change, high-flows and low-flows scenarios on the Meuse basin. AMICE – WP1 – Action 3 – technical report.

Foerster, H. & Lilliestam, J., 2010: Modeling thermoelectric power generation in view of climate change, Regional Environmental Change, pp. 327–338

Hall, J., Dawson, R., Sayers, P., Rosu, C., Chatterton, J.U. & Deakin, R. (2004): A methodologyfor national-scale flood risk assessment. In: Water & Maritime Engineering, Vol. 156, No. WM3, pp. 235–247. ISSN 1472-4561.

Kaplan, S. & Garrik, J.B. (1981): On the Quantitative Definition of Risk. In: Risk Analysis, Vol. 1, No. 1.

Sinaba et al. (2011): Quantification of the impacts of future floods on the economy in the international Meuse basin. AMICE – WP1 – Action 7 – technical report (unpublished).

Strobl, T. & Zunic, F., (2006): Wasserbau, Aktuelle Grundlagen – Neue Entwicklungen, Springer, Berline Heiderlberg New York.

Climate change and the impact on drinking water supply in the Meuse river basin

H. Römgens
RIWA Meuse, Maastricht, The Netherlands

ABSTRACT: The AMICE project shows we should expect a decrease of flow rates in the River Meuse of between 10% and 40% in low flow time periods. The total water demand from the River Meuse for drinking water production on average is 15 m^3/s while in the future longer periods of time shall occur with flow rates between 20 and 30 m^3/s at Monsin. In this situation there is a serious pressure on availability of the River Meuse as a source for the supply of drinking water, but not only due to quantitative aspects. Also the water quality can deteriorate to critical values due to lower flow rates caused by climate change and standards can be breached for longer periods of time.

1 INTRODUCTION

RIWA Meuse is an international association of water suppliers which use water from the River Meuse as a source for drinking water production. Six water suppliers are members: in Belgium Vivaqua (Brussels) and Water-link, a cooperation between Antwerp Water Works (AWW) and TMVW, and in The Netherlands *Waterleiding Maatschappij Limburg* (WML), Brabant Water, Evides (Rotterdam region) and Dunea (The Hague region). RIWA Meuse is not an Amice partner; in 2011 RIWA Meuse was asked to share some basic characteristics on water supply with the Amice project and has participated on the side-line of the project ever since. This article will first describe water supply in the Meuse river basin general and then the links with the AMICE sub study "*Quantification of the impacts of future low flows on the economy in the transnational Meuse basin.*" with the drinking water topic (Sinaba et al., 2010). Next the results of a study in The Netherlands called "*Effects of climate change on water quality at points of intake for water supply*" will be described.

2 ABSTRACTION OF WATER FROM THE RIVER MEUSE FOR WATER SUPPLY

Five water suppliers abstract water from the River Meuse at 7 locations as shown in Table 1.

Additionally WML has an abstraction point of river bank filtrated water at Roosteren (km 667, 2 Mm^3 per year). An overview of the river basin and the intake point is given in Figure 1.

The total abstraction in 2011 was 482 million cubic meters (Mm^3) with which approximately 6 million consumers in Belgium and The Netherlands were supplied with drinking water (Bannink, 2012). This averages out over the year to about 15 m^3 per second (m^3/s), but in summer conditions it has known to mount up to 21 m^3/s. Water abstracted from the River Meuse can be stored in reservoirs or the North sea dunes, after which it is purified to drinking water.

In time periods when there is not enough water flowing through the river or when the water quality is insufficient the storage in the reservoirs or dunes can be used or an alternative source, mostly groundwater, has to be activated. But at one location, at Tailfer, there is no storage nor is there an alternative source available. The maximum time periods that have to be bridged in dry,

Table 1. Abstraction of water from the River Meuse for drinking water production.

Location of intake	Km	River/tributary	Total volume abstracted in 2011 [Mm^3]	
Tailfer	520	Meuse	48	(Vivaqua)
Broechem	(600)	Albert Canal	56	(AWW)
Lier	(600)	Nete Canal	83	(AWW)
Heel	690	Lateral Canal	12	(WML)
Brakcl	(855)	Dammed Meuse	75	(Dunea)
Keizersveer	865	*Gat van de Kerksloot*	202	(Evides)
Scheelhoek	(915)	*Haringvliet**	6	(Evides)

*75% water from the River Rhine, 25% water from the River Meuse.

low flow periods or when water quality is insufficient are shown in Table 2. It shows high variation on different locations.

3 WATER QUALITY

The quality of the water flowing through the River Meuse has to be such that water suppliers can produce drinking water with natural purification technique. Article 7 sub 3 of the Water Framework Directive (WFD) even states that the level of purification treatment required in the production of drinking water should be reduced. RIWA Meuse distinguishes the following groups of water quality parameters:

- Traditional parameters such as Oxygen, Nitrogen and Phosphorus: these parameters usually meet the water quality standards and are no problem for the water supply.
- Pesticides: even though major improvements can be seen over the past decade we are still confronted with breeches of our water quality standard, 0.1 µg/L: for instance Isoproturon (IPU), Glyphosate and its metabolite Aminomethylphosphonic acid (AMPA), Chlorotolurone, Chloridazone, 2-methyl-4-chlorophenoxyacetic acid (MCPA) and Metolachlor (2011).
- Industrial substances: there are still some persisting problems with substances which keep breaching water quality standards and target values, such as Fluoride, Bromide, Diisopropyl ether (DIPE), Benzo(a)pyrene and Acetone (2011).
- Emerging substances: for these substances there are no water quality standards set yet but for which RIWA Meuse uses the target values from the Danube, Meuse and Rhine Memorandum (DMR), mostly 0.1 µg/L (IAWR et al., 2008): pharmaceuticals, X-ray contrast media, endocrine disrupting chemicals (EDCs), personal care and household products. In this category these substances breached the DMR-target values in 2011: Carbamazepine, Metoprolol, Sotalol, Ibuprofen, Metformin, Iohexol, Amidotrizoic acid, Iopromide, Ethylenediaminetetraacetic acid (EDTA) and Hexamethylenetetramine.

More information on substances which are relevant for drinking water production from the River Meuse can be found in Fisher et al. (2011).

4 CLIMATE CHANGE AS AN ISSUE FOR WATER SUPPLY

Climate change has an impact on water supply in dry periods. The flow rates of the River Meuse will be lower which leads to:

- Reduced availability of river water for various uses in time periods that coincide with an increase in demand by the various users;

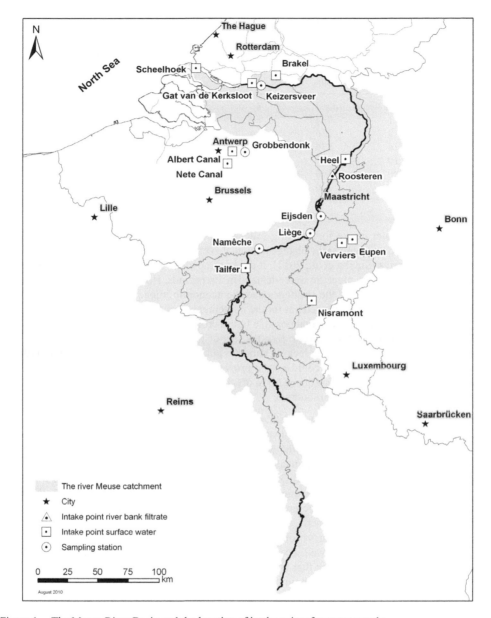

Figure 1.   The Meuse River Basin and the location of intake points for water supply.

- Deterioration of water quality due to reduced dilution of effluent of urban and industrial waste water treatment plants (WTTPs) by rain and groundwater (drainage).

4.1   *AMICE sub study "Quantification of the impacts of future low flows on the economy in the transnational Meuse basin."*

In the AMICE project there is an expectation for low flow conditions of 10% decrease in the time period 2021–2050 and 40% decrease in the time period 2071–2100. This concerns the MAM7 or 7-days mean minimum discharge. The AMICE sub study *"Quantification of the impacts of future low flows on the economy in the transnational Meuse basin."* did not succeed in calculating the

Table 2. Bridging periods and alternatives in situations where water from the River Meuse is unavailable.

| Location | Maximum bridging period | | Alternatives available | |
	days	hours	source	water supply
Tailfer	n.a.	n.a.	no (not necessary)	no (not necessary)
Broechem	31	744	no	yes
Lier	9	216	no	yes
Heel	14	336	yes (groundwater)	
Brakel	10	240	yes (groundwater)	possible (no contracts)
Keizersveer	90	2160	no (not necessary)	yes
Scheelhoek	30	720	yes (groundwater, limited)	yes (limited)

consequences for water intake for drinking water production because for longer periods of time changes were only tested against the change of annual averages and not against the change in daily or monthly time series (Sinaba et al., 2012). It did become clear that bridging periods and storage capacities are too limited to meet the water demand in dry periods. However, data to make this more tangible as well as the size of the shortages and the economic impacts are lacking. A system of reliability assessment was introduced but for the implementation of this system the necessary data was lacking. The end conclusion is that possible future drought conditions impose additional stress on water supply systems, which have to face the risk of frequent water shortages and significant economic impacts. However, due to lacking data, the quantification of future water shortages and the estimation of its frequencies could not be provided. There is still a demand on research to answer the question what the consequences are of a flow rate reduction of 10% to 40% in dry, low flow periods for drinking water production:

- Based on current statistics the low flow MAM7 once every 30 years will reach at Chooz approximately 13 m^3/s and at Monsin approximately 28 m^3/s. If climate change will lead to a reduction of 10% to 40%, say average 25%, the MAM7 flow rate at Monsin will reach approximately 21 m^3/s (17–25). Therefore it is thinkable that during a longer period of time the flow rate at Monsin will be in the region of 20 to 30 m^3/s. This flow then has to be divided over the Albert Canal and the Common Meuse. Compared to the average capacity needed by only the water suppliers of 15 m^3/s it is clear that tension could arise between different users because the total water demand will be even higher.
- In the sub study the impact for various sectors, such as energy production, agriculture, water supply and navigation, is reviewed separately. However, the low flow in dry time periods, which decreases the amount of water that is available, coincides with increasing water needs (for cooling, drinking etc.) in that same time period. This summation issue deserves more attention, perhaps a follow up study.
- In the sub study the availability of water from the River Meuse in dry periods of time is addressed as an issue. However, availability is not the only issue. The other side of the same medal is water quality. Because in a time period of 10% to 40% reduction of low flow leads to a corresponding deterioration of water quality, forcing water suppliers to cease their intake more often, sooner and/or longer. This increases the vulnerability of drinking water supply.

4.2 Effects of climate change on water quality at intake points for drinking water production

In a study by the Dutch National Institute for Public Health and the Environment (RIVM) and Deltares the water quality changes of the Rivers Rhine and Meuse are calculated for various years for different climate scenarios from the Dutch Delta Program (Wuijts et al., 2012). This has been done for an average year (1967), a dry year (1989, once every 5 years) and very dry year (1976, every 80 years). The climate change scenario W$_{plus}$ from the Royal Netherlands Meteorological Institute

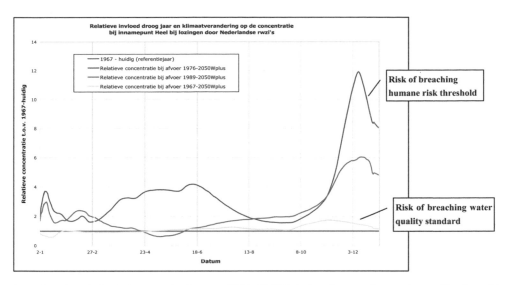

Figure 2.   The relative influence of a dry year (1989–2050Wplus) and a very dry year in combination with climate change (1976–2050Wplus) in comparison to an average year without climate change at the Heel intake point by emissions of Dutch waste water treatment plants (From: Wuijts et al., 2012).

(KNMI) was used as a worst case scenario. In this study two effects are described separately: emissions from upstream countries (transboundary loads) and emissions from WWTPs in The Netherlands. The study focusses on two groups of substances:

- Substances for which water quality standards are set in The Netherlands for surface water that is used for drinking water production. Current substances of concern are pesticides for which the standard is 0.1 µg/L. Several pesticides are already frequently breeching or touching this water quality standard.
- Emerging substances for which there are no standards for surface water that is used for drinking water production in The Netherlands, for instance pharmaceuticals (see chapter 3). One substance, Carbamazepine, has been used as an example as it is already frequently found at or above the DMR-target value of 0.1 µg/L. The lowest reported human risk threshold is 1 µg/L.

On this basis the following hypotheses were tested:

- If the concentration of pesticides increases with a factor 2 will this lead to an increase of breaching incidents?
- If the concentration of Carbamazepine increases with a factor 10 will the humane risk threshold be breached?

It was calculated how much the concentration of these substances was increased due to the decrease in flow rate and how many days the standard or humane risk threshold would be breached. The calculations did not take any degradation of the substances into account but only the dilution effect. Some results are shown in Figures 2 and 3.

Figure 2 shows what the relative influence is of climate change in a dry (1989) and a very dry year (1976) at the Heel intake point. During longer periods of time concentrations are a factor 2 or more higher. For instance, in a dry year the concentration is a factor 2 or more higher as of September and approximately 100 days there is a serious risk of breaching standards for pesticides. In a very dry year the expected concentration is a factor 10 times higher as of December and during approximately two weeks the humane risk threshold for a pharmaceutical is breached.

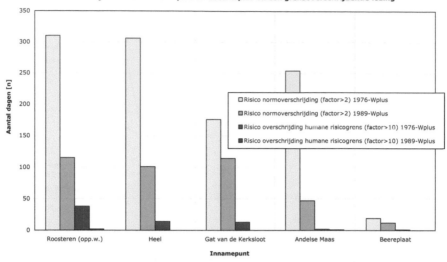

Figure 3. Number of days per year with a risk of breaching the water quality standard or the humane risk threshold at intake points in the River Meuse basin for various scenarios by transboundary emissions (From: Wuijts et al., 2012).

Table 3. Calculated number of days and the longest continuous period with risk of breaching water quality standards at intake points in a dry year and fast climate change (1989–2050Wplus) (Modified from: Wuijts et al., 2012).

| | Risk of breaching standards in a dry year (1989) [days] | | | | |
| | Transboundary | | Dutch WWTPs | | Current bridging capacity [days] |
Intake point	Total number	Longest Period	Total number	Longest period	
Roosteren	115	85	121	86	Bank filtration
Heel	101	76	128	95	21–120
Gat van de Kerksloot	114	47	234	179	60–90
Brakel	47	17	103	103	10–28

Figure 3 shows the effect of transboundary emissions on the number of days water quality standard of humane risk threshold is breached at the various intake points in the Dutch part of the River Meuse Basin.

The general conclusion is that due to climate change the flow rate of the River Meuse in dry time periods will be so much lower that concentrations of substances increase 2 to 10 fold. Structurally longer periods of time can occur in which water quality standards are breached for the current substances of concern. Table 3 shows this for the dry year 1989, which has a frequency of occurrence of every five years.

The consequence of the results as shown in Table 3 is that water suppliers cannot use water from the River Meuse in these periods of time and have to rely on their storage – if any – or alternative resources. The last column of Table 3 shows the current bridging capacity at the intake points, most of which are inadequate. For a very dry year (1976) this will only be worse.

5 CONCLUSIONS AND RECOMMENDATIONS

5.1 Conclusions

- The AMICE project shows we should expect a decrease of flow rates in the River Meuse of between 10% and 40% in low flow time periods. As a result AMICE states that future drought situations shall lead to frequent periods of water shortages and a significant economic impact (which unfortunately has not been quantified in the project).
- The total water demand from the River Meuse for drinking water production on average is 15 m^3/s. In the future longer periods of time with flow rates between 20 and 30 m^3/s at Monsin shall occur. Only this fact makes clear how pressing the situation can get for the supply of drinking water and the other water users.
- There is no overall picture of the water demand of all sectors of use against the available water in the River Meuse during dry and very dry years.
- The RIVM/Deltares study show that water quality can deteriorate to critical values due to climate change and standards can be breached for longer periods of time (Wuijts et al., 2012). Water quality issues have not played a significant role in climate change studies up until now, AMICE has focused only on quantitative issues.

5.2 Recommendations

The AMICE project should get a follow up in which the following aspects from the perspective of water supply and other users must be studied:

- From the sub study 'Impacts of future lower flows': "Identifying water shortage periods, where the water demand exceeds the water availability enables the water supply operators to develop adaptation strategies". We recommend to link this to the identification of the total water demand from all sectors in dry time periods, especially during extreme low flow conditions.
- To determine the consequences of lower flow rates for the water quality in the River Meuse in relation to the water quality standards and target values for surface water used for drinking water production.
- Incorporate water quality in adaptation strategies for climate change and estimate which measures need to be taken to reduce emissions in such a manner that even in future low flow conditions water from the River Meuse is suitable as source for drinking water production.
- Also the water quality issue should be taken into account in the development of policies concerning climate change and vice versa should climate change become an issue in the River Basin Management Plans under the Water Framework Directive.

REFERENCES

Bannink, A.D. 2012. Annual Report 2011. Meuse (In Dutch, *Jaarrapport 2011. Maas*). Maastricht: RIWA Meuse.

Fischer, A., Bannink, A. & Houtman, C.J. 2011. Relevant substances for Drinking Water production from the river Meuse. An update of selection criteria and substances list. Haarlem/Maastricht: Het Waterlaboratorium/RIWA Meuse.

IAWR (*Internationale Arbeitsgemeinschaft der Wasserwerke im Rheineinzugsgebiet*), IAWD (*Internationale Arbeitsgemeinschaft der Wasserwerke im Donaueinzugsgebiet*) & RIWA Meuse. 2008. Danube, Meuse and Rhine Memorandum 2008. Cologne/Vienna/Werkendam.

Sinaba, B., Bauwens, A., Buiteveld, H., Brede, R., Deckers, P., Degré, A., De Keizer, O., Detrembleur, S., Dewals, B., Fournier, M., Guilmin, E., Hissel, F., Huber, N., Kufeld, M., Van der Mark, R., Vanneuville, W., Pirotton, M. & Schüttrumpf, H. 2012. Quantification of the impacts of future low flows on the economy in the transnational Meuse basin. Charleville-Mezieres: EPAMA/AMICE.

Wuijts, S. (RIVM), Bak-Eijsberg, C.I. (Deltares), Van Velzen, E.H. (Deltares) & Van der Aa, N.G.F.M. (RIVM) 2012. Effects of climate change on the water quality at intake points for drinking water. Analysis of substance calculations. (In Dutch, *Effecten klimaatontwikkeling op de waterkwaliteit bij innamepunten voor drinkwater. Analyse van stofberekeningen*). De Bilt: RIVM Report 609716004/ 2012.

Transboundary Water Management in a Changing Climate – Dewals & Fournier (Eds)
© 2013 Taylor & Francis Group, London, ISBN 978-1-138-00039-1

Natural water retention, a no-regret measure against future water-related risks and an opportunity for local communication

M. Fournier
EPAMA, Charleville-Mézières, France

M. Lejeune
RIOU, Hasselt, Belgium

P. van Iersel & R. Lambregts
Waterboard Brabantse Delta, Breda, The Netherlands

C. Raskin
Commune de Hotton, Belgium

ABSTRACT: Three AMICE pilot investments are using natural water retention as a means to adapt to future climate conditions and associated water-related risks. The benefits of this technique are highlighted and call for the creation of a network of similar local projects on the Meuse catchment. Experiences on communication towards the local population are also described, as they are an integral part of the success and promotion of these measures.

RESUME: Trois investissements pilotes d'AMICE apportent des solutions pour s'adapter aux conditions climatiques futures et aux risques hydrauliques associés via la rétention naturelle de l'eau. Les bénéfices de cette technique sont mis en avant et appellent à la création d'un réseau de projets similaires sur tout le bassin de la Meuse. Les expériences des Partenaires pour la communication vers les populations locales sont également décrites, car elles sont partie intégrante du succès et de la diffusion de ces mesures.

1 INTRODUCTION

In the future there will be more floods and more droughts. Whatever we do now, we cannot stop climate change. Adaptation to changing circumstances is a necessity – but we can choose how we take action. AMICE gives us the opportunity to test all the options and build them into one overall strategy. The river is bound to respond in ways we know but we are also anticipating some surprises. Water is the 21st century's essential resource. Improvements in natural water retention (NWR) can often be achieved through low impact, small-scale land-use changes. Wetlands have a role in buffering of water discharge variations. Conservation of the water retention areas is a good way of combining climate adaptation, sustainable development and involvement of local communities.

The goal of AMICE workgroup 2 is to promote the network of (NWR) areas throughout the Meuse basin as a solution to adapt to climate change. Three pilot investments have been developed in the framework of AMICE. They are located in Sankt-Vith and Hotton (BE) and Steenbergen (NL). The Partners responsible for their implementation are respectively RIOU/BNVS, the community of Hotton and the Waterboard Brabantse Delta.

Partners held 5 meetings between 2009 and 2012 to share their experience about project development and implementation, taking climate evolution into account, and involving the local population. Site visits were organized all along the projects' developments to understand better the local threats and solutions. Site visits were also useful for the river managers to know each other better and

Figure 1.   Natural bogs can stock huge amounts of water.

strengthen cooperation. The visits were organized in Sankt-Vith (01.10.2010), Hotton (28.04.2009; 21.06.2012) and Steenbergen (18.03.2010; 30.09.2011).

2   DESCRIPTION OF INVESTMENTS

2.1   *Climate buffers on the Amblève, near Sankt-Vith*

On and near the plateau of the "Hautes Fagnes", floods occur when the soil is saturated and heavy rains cannot be absorbed. Floods must be regarded as a problem on a river basin scale. One way to approach this, is to store the rain where it falls. In this way, the peak can be leveled.

River valleys are brought back to a natural state in order to keep the water longer in the soil and create natural buffers to cope with the effect of climate variability. Originally focused on flood prevention, it turned out to have a positive effect on the restoration of the natural vegetation and on the reduction of low-flows during summer.

The goal is also to set a monitoring program based on hydrological and weather measurements. Additional botanical and entomological monitoring was carried out to be added to five years of data already available from three sites studied before. It helps to quantify which volumes are stored thanks to the ecosystem. People are convinced quicker of the efficiency of changing the land use. But the experience from the local organizations and similar measures carried out elsewhere for more than 20 years are also proof of this aspect. This is a no-regret measure that is good both for floods and droughts.

But there is no effect in Rotterdam, of course. The first hydrological measurements are being generated but more years are needed to demonstrate a significant contribution. Even if the effect is

46

only local, such projects are quite easy to reproduce elsewhere and then could have a more general impact.

Moreover, it sets an example for the integration of the Water Framework Directive, Natura 2000 and the Floods Directive. Natural restoration leads to enhanced ecological status and to natural floods mitigation. Effects are furthermore visible in the attractiveness of the region. The number of tourists cycling in the area on the RAVEL network has increased thanks to the renaturation.

2.2   *Integrated management plan of the Naives basin, near Hotton*

Ny is one of the most beautiful villages in Wallonia. A typical threat to it is the Naives river flowing underneath the village. It is a relatively small catchment area, but it is characterized by very steep slopes in a hilly environment. These slopes create flood problems on a very regular basis. The rainwater couldn't go anywhere with the water level in the underground vault limited to a mere 30 cm.

Moreover, the waters of the Naives were of poor quality due to an old sewage system. The situation worsens during low-flows that are really severe in such a small basin. An integrated approach was needed.

The goals is to combine flood prevention in Ny and in the downstream area of Melreux with a better water quality, limitation of toxic blue-green algae, sustainable fishing activity, conservation of the natural resources, improvement of the architectural value and tourism.

Fight against extreme floods
Part of the solution is to clean and enlarge the vaulting of the Naives across the village to evacuate its waters quicker. To compensate this flow acceleration, a dike is built downstream of the village of Ny to slow it down again. The Natura 2000 area is becoming a temporary storage for 160 000 m³ of water. The dike is built along the road, so the river plain can function as a storage area, thus avoiding flooding of the downstream village of Melreux. A culvert of 23 m width enables the evacuation of the water. The calibration is based on the 1000 years return period. The outlet was also designed in order to let the fishes swim through. Poles are planted upstream of the outlet to prevent debris damming. The floodable area can still be used for grazing and the dike itself will be used as a cycle trail, part of the RAVEL network. A special entrance was made to help the farmers to use the area.

Fight against extreme low-flows
Part of the solution is to clean the riverbed and improve the tree line. Works at the village consist of cleaning the underground part of the river, separating waste water and clean water in different sewer systems and putting electric wires underground.

One of the additional problems to be dealt with in the river Naives is blooming of blue-green algae. This only occurs in dry periods when discharges are extremely low and the water stagnates due to the presence of dirt dams. The algae make the water look very unattractive, but this is hardly the biggest problem. In fact some of those algae are toxic ; they make the water poisonous for other organisms, included men and thus have a very bad influence on the water quality. The river has been tidied up now and trees were planted. In two years time, the water quality hugely improved and the system looks as a real river again.

The technical components of this project can be copied to be used in comparable regions. Since it has a small-scale effect, if more measures would be implemented, a significant change for the Meuse could be reached. Moreover, a local project like this can contribute to the sense of solidarity and upstream-downstream relations.

2.3   *Steenbergse Vliet*

The Steenbergse Vliet is a small river, with a typical distinction in differential head between the Belgian side and the Dutch side. The Dutch part of the river hardly has a slope over a stretch of 30 kilometers. Retention times are long. Upstream, the slope of the river is steeper. The Steenbergse Vliet discharges into the Volkerak Zoom-lake. This is a heavily modified lake that used to be part

Figure 2. Location of the new dike and ponds on the Naives plain.

of an estuary. Its water level depends on the management procedures related to climate change adaptation plans.

Both upstream and downstream situations are threatened by climate change. Downstream, the sea level can be increased up to 50 cm. If a high water discharge is running down from the Meuse, this leads to a serious challenge for mitigation of the flood risk.

An area of about 8 ha is being changed from agriculture usage into nature. The whole area is part of the major bed of the river. After the completion of the project, nature can go its course and the costs for the water board in times of flooding will decrease significantly. A wetland has been recreated, which is always connected to the river and has a fluctuating water level. It has different goals: water retention, ecological connection zone and as fish breeding area. A bat wintering place was made by re-using concrete culvert materials. Tourism is not forgotten, with the creation of a cycle trail and a GPS tour.

The project has a very broad scope as stated above: risk reduction, nature development and tourism are all intended results. In addition, agricultural goals were not hampered by this project. It's a no-regret action.

Climate change will generate more floods in the Meuse as well as a higher sea level, which will result in a maximum elevation of the Volkerak lake of 2.3 m for periods of 2–3 days. In this situation, there is no more space below the bridges for the ships to pass. Salt intrusion can also become a threat in the future. The lake is used for flood storage with a return period of once in a 1400 years; but with climate change it would be once every 550 years in 2050. The Steenbergse Vliet will help store enough water to cope with a water rise in the river of 1.2 m.

48

Figure 3.   The Project in June 2012 seen from the Steenbergseweg.

Stopping any new constructions in the watercourse will also help reduce the amount of damage.

3   SHARING EXPERIENCES ON LOCAL COMMUNICATION

Partners shared experiences on project development, wetland restoration and river management. More interesting however were the discussions on how to communicate with the local landowners, the mayors and inhabitants of the area. This communication served multiple purposes:

– explain the changes in the land use,
– explain the consequences of climate change,
– develop a basin's solidarity, the feeling of belonging to a common territory and the consciousness to share the same resource

Here are some examples of communication activities related to nature conservation and river basins that were presented or developed within this workgroup.

3.1   *The river pearl mussel game*

BNVS exposes the example of a game called the "river pearl mussel". It teaches children the biology of the mussel as well as good practices to protect rivers. Children move from one square to another answering quiz questions and throwing the dice. When they win they collect a mussel ring. The winner as a complete adult mussel. The drawings on the board represent the river ecosystem.

3.2   *Sensitization action in the Ardennes*

In September 2011, RIOU's local partner Natagora/BNVS carried out a rather playful sensitization action on natural water retention. With this action, people got better acquainted with the

Figure 4.   Playing around the concept of natural water retention.

organisation, but above all, it heightened the public awareness of the 'natural water retention' theme which is an important main line in the nature reserves.

A postcard was sent to 32.000 households in the German-speaking region of Belgium. On its back side, is a small quiz: people could go to the website and answer a short (and easy) question. About 5000 people participated.

Winners got rubber boots, backpacks, river pearl mussel games and free BNVS-memberships.

3.3   *Holding public meetings to inform the population in Ny*

A public meeting was organized on December 10th, 2011 to inform the population in details about the works and the schedule for 2012. Because the flood dike in the Naives meadows is only the top of the iceberg! The integrated management of the Naives basin includes a flood control dike, maintenance of the river to reduce risks related to low-flows, improved waste-water collection and treatment, renovation of the underground river vaults, improvement of the village's architecture (among others: lightening of the fountains and underground electric wires). 50 inhabitants of Ny came to the meeting. There were many practical questions but no opposition, which proves the success of the past years' negotiations and the high quality of the planning.

3.4   *The GPS route around the Steenbergse Vliet*

A bike trail has been traced along the Steenbergse Vliet and passes the AMICE pilot investment. Regularly, the visitor comes across a Point of Interest (POI) and can access information. To have the information of the POI, you need at least a Smartphone or I-phone with internet and GPS equipped.

Which APPS you need can be found on http://www.kunstroutesbrabant.nl/Kunstroutes-bkkc/ Het-land-van-contrasten.aspx in Dutch.

For example, there is an interactive video about the project area used in combination with a MP3 player and GPS presenting:

– cultural history: water defence lines, fortresses, inundation areas, historical sluices, floodings from the sea in the past, future usage of the Volkerak Zoommeer, the typical open landscape, dike landscape and other historical developments since 1930;
– water management, in the past , now and in the future;
– history of the town of Steenbergen: harbour, fortress town, salt and peat exploration.

Figure 5. Explaining the land use changes in a modern way.

4 CONCLUSIONS

This AMICE workgroup further analyses the opportunities offered by Natural Water Retention to tackle floods and low-flows. The show cases developed here are very different from one another, yet they use the same principles: make room for the river, transform agricultural land into natural space and conserve Natura 2000 zones within floodplains.

Transnationality is reached through the transferability of these investments and publicity done. It also contributes to the international solidarity effort to adapt to the consequences of climate evolutions, demonstrating that all tributaries can have a role to play. If a more greater proportion of the river valleys in the catchment area can be used as NWR zones, the influence on the whole of the Meuse basin will become quite significant. Moreover, these pilot investments can be qualified as no-regret measures and are ecologically efficient. They improve biodiversity at local level.

Communication examples demonstrate that fear is not a good way of involving people in adaptation to climate change. It is much better to try and show actions the target public can do. It is also important to have people understand that climate is already changing and that it is not only a far future threat. Finally, the message is that solutions do exist and can be implemented right now. Solutions bring several other advantages and adapting to water-related risks is also an opportunity to improve our way of life!

Structural protection against future water-related risks and solidarity among the Meuse countries

M. Fournier
EPAMA, Charleville-Mézières, France

J. De Bijl
Waterboard Aa en Maas, 's-Hertogenbosch, The Netherlands

G. Demny & C. Homann
Wasserverband Eifel Rur, Duren, Germany

K. Maeghe
nv De Scheepvaart, Hasselt, Belgium

M. Linsen
Rijkswaterstaat, Lelystad, The Netherlands

ABSTRACT: There are many flood-water management constructions operating already in the Meuse river basin and more are planned in the next decades. Herein lie some big challenges. How to design new water management structures that are able to deal comprehensively with flooding, drought and increasing water demand. How to adapt existing flood control measures to cope with ever more extreme events. Through AMICE, new approaches to these challenges have been tested by three highly innovative projects in Germany, Flanders and the Netherlands.

RESUME: Il y a de nombreux ouvrages de gestion des inondations en fonctionnement dans le basin versant de la Meuse, et d'autres sont en projets pour les prochaines décennies. Il s'agit ici d'un grand challenge. Comment dimensionner les nouvelles infrastructures de gestion de l'eau pour qu'elles puissent répondre à la fois aux inondations, aux étiages et à une demande en eau croissante ? Comment adapter les mesures existantes de contrôle des crues pour faire face à des évènements plus extrêmes ? Dans le cadre d'AMICE, de nouvelles approches ont été testées à travers trois projets innovants en Allemagne, en Flandres et aux Pays-Bas.

1 INTRODUCTION

There are major water infrastructures in the Meuse basin, f.e. hydropower dams in Germany, large and flat water storages in the Netherlands, channels and locks in Belgium. The impacts of climate change on the management of these infrastructures are yet not clear enough and Partners have studied how to deal with this uncertainty.

It also demonstrated how the adaptation of water management could contribute to the reduction of impacts of high and low flows. It soon appeared that transnational cooperation is compulsory to avoid drawbacks, share efforts and exchange useful knowledge. The objective was to design and implement measures against low-flows and floods risks with transnational benefit at the scale of the international river basin.

This issue has been shared with partners from other countries and other sectors in order to improve the technique and integrate the investments with other works being carried out in the basin. Site visits were organised to experience new areas of the Meuse basin and improve understanding of

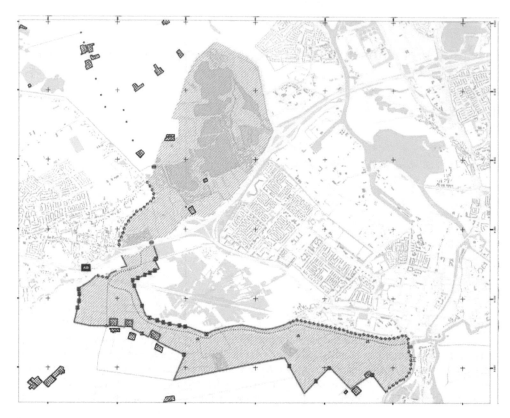

Figure 1.   Map of the HOWABO reservoir.

local constraints and issues: HOWABO (19.03.2010 & 04.09.2012), Rur (06.04.2011) and Lock of Ham (27.09.2012).

2   DESCRIPTION OF INVESTMENTS

2.1   *HOWABO (Hoog Water 's-Hertogenbosch)*

There is a need to compensate negative flood effects due to an acceleration of the flood wave resulting from defensive flood measures upstream at the Grensmaas. The peak discharges of the tributaries can reach the Meuse simultaneously and generates floods in 's-Hertogenbosch. An area of 750 hectares is adapted to become a water retention area. This is designed to be flooded once every 100–150 years. The peak discharges can be shaved, and the water has room to continue its course.

The flood reservoir is a multifunctional area where one can maintain agriculture, nature conservation and recreation. An inlet construction lets in water from the Drongelens Canal into the Vughtse Gement. Water retention has been combined with recreation purposes and with development of nature. NGO's have been involved. They were made part of a steering group of decision makers. In addition, they are part of the project group that manages the water retention area. They named the area the "green river". The water quality is also improved in the area.

The design of the water retention inlet was based on the actual hydraulical conditions including Maaswerken (3 million m³). The project, however, is flexible and adapted to conditions affected

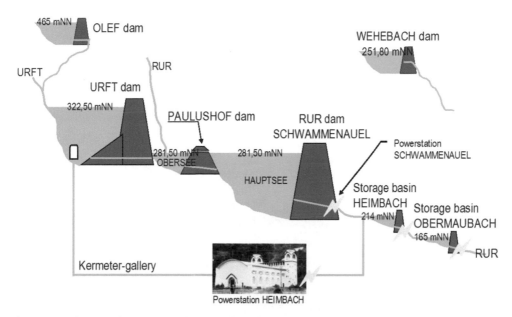

Figure 2. Scheme of the Rur reservoirs network.

by potential climate change scenarios (5 million m³). This amount has been established in two principles that are – according to the Partners – crucial for the success of a project such as HOWABO:

– solidarity (no shifting of problems to other water managers/public downstream),
– uncertainty about the increase of river discharges.

Therefore the project is an example for cooperation and adaptation in the river basin of the Meuse. Either way, shifting the incidental peak discharges will be prevented.

2.2 Rur's reservoirs network

The Rur is one of the largest tributary to the Meuse. It is controlled by a total of 6 reservoirs having a total storage volume of about 300 million m³. The reservoirs are operated for drinking water production (about 700,000 people depend on it), low-water enrichment, mitigating floods, cooling water for the industry and tourism. The mean flow into the Meuse is reduced by 86 m³/s and the low flow is increased up to 13 m³/s thanks to the reservoirs. Impacts of climate change have to be considered in the future flow control system to avoid an increase in damages. Moreover, the reservoir control system was designed on the basis of observed time series over the last 100 years but does not account for future changes.

A model has been set-up in order to assess the hydrological and hydraulic system in the Rur basin. This requires a detailed modelling from the most upstream reservoir down to the inflow into the Meuse. Thanks to this new tool, it becomes possible to investigate floods and low-flows. Two situations are envisaged: for the present climate and for the future climate. The reservoir control system had to be adapted based on the AMICE climate scenarios as well as water consumption trends in the downstream Netherlands.

The scenarios can be used to assess climate change influences on dry periods and floods.

Downstream hydrological models have been combined with risk analyses. In this way, damages can be projected for flood scenarios. Combining these insights, it becomes clearer how a peak discharge moves through the system, and which damages it might cause. Parameters are continuously adapted to the weather. Based on weather forecasts, reservoir reactions are predicted. This allows for peak discharges to be "accommodated." The operational rules have to be more flexible

Figure 3. Building the pumping station on the Lock of Ham.

though. The maximum capacity of the Rur is now known. The flood scenarios show which type of measures are necessary when the maximum return period capacity will be exceeded. Options such as retention areas become necessary: this necessity is easy to communicate thanks to insights in risk, flow regime and peak discharges.

The paper industry has the priority in case of water shortages. On the other side, hydropower is only produced when there is enough water. Nowadays the paper industry sometimes has to stop for 2 months per year.

2.3 Lock of Ham, Albert Canal

The Albert Canal runs from the inland port of Liège to the port of Antwerp and is fed entirely by the Meuse. The total fall is 55.5 meters. Inland navigation accounts for the transport of about 40 million tons per year. This can be compared to a daily train of 6000 trucks per day. The canal serves as navigation route and as water resource for various purposes (irrigation, drinking water production, nature, process water for industry, cooling water for the production of electricity), as well as a migration bypass for many fish species.

The necessary discharge of 25 m³/s is easily met for most of the year, as the average Meuse discharge is about 200–250 m³/s. In dry periods of several weeks, however, the Meuse discharges can drop below 45 m³/s. Based on the Meuse Discharge Treaty, Flanders and the Netherlands divide the available water during droughts, with always a guaranteed discharge for the non navigable Common Meuse, which makes water shortage measures necessary. Climate change may cause a more frequent occurrence of low Meuse discharges. Frequent droughts also lead to increased water demands in Flanders. The water shortage problem is known for years. Climate change just makes it worse.

Pumps have been installed on the Lock of Ham as a pilot case. They have a capacity of 15 m³/s and are designed in a fish-friendly way. In periods with sufficient discharge of the Meuse, these pumps work in reverse direction as a hydro electric power station. The large fall at the lock (10 m) can produce green electricity for about 2000 inhabitants and contribute to the production of renewable energy.

The Lock of Ham is the first to be equipped on the Canal because it has the biggest water consumption. The idea of Archimedes screws for pumping water back at a lock is not new. But it is the first time that it is realised on a canal as long as the Albert Canal and with locks so high. The installation on each lock costs about 7 million €.

nv De Scheepvaart will use the electricity for the exploitation of the lock and sell the remaining part from the hydropower station. An economical cost-benefit analysis was carried out and the project will indeed be cost-effective. The maintenance and operation has also been included in the evaluation. The return on investment for the adjustment as water power plant will be possible after 7 or 8 years.

3 SHARING EXPERIENCE ON THE BENEFITS OF SOLIDARITY

The Meuse Treaty signed between Flanders and the Netherlands is also an example of international agreement to share the water. Especially during low-flows, it is important to agree on the priorities given to navigation, drinking water or minimum ecological water levels. But when river discharges are much too low, some needs cannot be met, leading to economical consequences. With climate change, these situations are bound to increase. To respect their part of the Treaty, Flanders is ready to make huge investments and install pumps on all the locks of the Albert Canal to save water and to reduce the bottleneck of the actual lock for fish passages.

The dimensioning of the HOWABO reservoir took into account many parameters. Though it is located on the Aa river, a tributary to the Meuse, works on the common Meuse in the Province of Limburg and their impacts on discharges have been included in the calculations. A thorough knowledge of the upstream region is the basis for all large projects. Working together on river basin level is cost-efficient!

L. Verheijen, Dijkgraaf of the Waterboard Aa en Maas, acknowledges the need for international collaboration along the river Meuse[1]: "People in the upstream area don't know the downstream part and in the downstream area people know almost nothing about what's going on upstream. People in The Netherlands, downstream, always tend to think that Belgium and France should do more about water quality and water retention. It is easily forgotten that these countries also can ask something from The Netherlands. Fish migration for instance is an important problem that can largely be solved by The Netherlands."

Decision-making at the Wasserverband Eifel-Rur is also an open process. Policy-makers and stakeholders along the river are represented in the Board. For the AMICE action, a Joint Expert Group has been created, involving the Waterboard Roer en Overmaas (the Netherlands). Professor W. Firk, Chairman of the Waterboard Eifel Rur, explains the importance of involving the neighbors[2]: "We have a well established collaboration and knowledge transfer concerning the management of the river Rur. From my point of view, widening the collaboration to the whole river Meuse is the consequent next step. Though we have already a cooperation on state level by the International Meuse Commission, we furthermore should strengthen the cooperation by the water managers on site."

Though there is no drought problem on the Rur, the river managers are conscious that they have to think about the needs of their downstream neighbours, especially navigation on the Meuse's canals.

4 CONCLUSION

AMICE's workgroup n°3 demonstrated how quantitative water management can contribute to the reduction of impacts of high and low-flows. The partners involved were also interested in knowing how to adapt their systems to the future hydrological context. Flexibility of the infrastructures like around 's-Hertogenbosch is to be promoted as a response to the still uncertain magnitude of climate change.

These measures are of huge dimensions and their management will have a noticeable impact on other regions of the transnational basin.

As regards adaptation to climate change, these measures have positive 'spill-over effects': the modification of the management rules of the Rur reservoirs system will guarantee water provision to all users (drinking water production, hydroelectricity, cooling water for industry, leisure) in the future decades. They are often called 'win-win' measures as all the water users find something positive in the project: the installation of fish-friendly pumps on the Albert canal to reduce water consumption on sluices will improve both navigation on the canal and discharges on the Meuse river during drought spells.

REFERENCES

[1]Meuse and Climate, AMICE Newsletter n°2, June 2010.
[2]Meuse and Climate, AMICE Newsletter n°3, December 2010.

Transboundary Water Management in a Changing Climate – Dewals & Fournier (Eds)
© 2013 Taylor & Francis Group, London, ISBN 978-1-138-00039-1

Crisis management: the AMICE exercise in November 2011

B. Tonnelier
Etat Major de Zone Est, Metz, France

Epinal the 7th, Bar le Duc the 8th and 9th, Charleville-Mézières the 10th, and finally November 17th and 18th in Metz. The month of November 2011 has seen 4 operational centres, 3 at district level and 1 at zone level, deal with a major flood of the Meuse in the framework of a common exercise.

The common exercise was not simultaneous: in order to evaluate its scope better, to ensure a better collaboration of all and to facilitate the moderation by a unique team, a sequential play was required.

Each operational district centre could indeed benefit of a specific setting aimed at its own stakes and problems; just as the zone level could focus on its missions of support and coordination.

Explanation on this topic

The territorial organisation of crisis management, in France and under State governance, is composed of 3 levels of competence: district, zone and national. The prefect of the district hit by the incident is regarded as the director of rescue operations and the coordinator of State services in the district, as long as the crisis or its consequences do not reach beyond the district boundaries. If it becomes wider, the prefect of the defence zone must be able to proceed with action and coordinate the district prefects concerned. In all circumstances, he has to support them by making means from the Zone available, either human or material. The chain of responsibilities then moves up to the national level and the involvement of political decision.

As far as territorial communities are concerned, the mayors do indeed play a predominant role, but they are also supported by the district level administration in the achievement of their public missions and roads safety.

In France, the river Meuse flows through 4 districts, and even more if one counts the tributaries. The Eastern zone for defence and security thus appears the best level to develop this proposition of exercise and to support EPAMA.

The involvement of the district services was achieved through the district prefectures. The knowledge of the land, plans and procedures has naturally enriched the global scenario, based on common objectives with, in the background, a centennial flood increased by +15%:

– evacuations and emergency
– protection of goods
– socio-economic impacts
– maintenance of infrastructures
– continuity of services
– crisis communication
– civil-military cooperation.

On these bases, the scenario-writing teams have elaborated, from the defence zone to the district, and from the district to the local community, events and incidents that required appropriate reactions, a fast coordination and research of means, as well as a constant dialogue between the stakeholders of crisis management.

More precisely, this implied to face many challenges, such as:

– provide assistance and bring to safety the threatened populations, especially the weakest (hospital in Charleville-Mézières, in Saint-Mihiel, elderly homes . . .),

Figure 1. Etat Major de Zone Est.

- protect or substitute the vital networks (water, communication, electricity . . .),
- protect the cultural heritage or the infrastructures (bridges on the Meuse, dike at Dun Sur Meuse . . .),
- check on important economic tools (PSA factory in Villers-Semeuse, nuclear power station of Chooz . . .),
- maintain information to the public (hotlines, press releases) and organise travels of political persons.

All this has to be done while involving all services of the State and the local communities (fire brigades, police, civil defence, equipment, health, housing, education), electricity and gas companies, the armies supporting the civil authorities, but also private stakeholders (travel companies, ambulance, public works . . .).

Of course, transboundary cooperation was not forgotten, through exchanges between the Ardennes district and the province of Namur, the zone operational centre and the regional crisis centre in Wallonia.

Primarily designed as a strategic play, including high-tech tools (website Vigicrues, maps and animations by SERTIT in Strasbourg), the actions on field have not been excluded. For example, the strengthening of the dike in Dun Sur Meuse by a regiment of the French Army.

In addition to testing some decisions from the crisis cells, these actions offered the press the opportunity to cover the event, to have a realistic view on the involvement of participants, as well as the difficulty and usefulness of the exercise.

Creating a link and dialogue between the emergency professionals, training them to taking advice and deciding before action, test the flaws in the plans, are renowned constants of the civil security exercises. Either held indoor or outdoor.

Drawing lessons is a must. AMICE led everywhere to return of experiences. Each level of responsibility has drawn its critical analysis, on processes and implementation. Without getting into details, it can be mentioned that problems appeared but also solutions. Some of them are still to be further explored, including at the national level, always eager for recommendations.

The dominant feeling is satisfaction. Doubtless the satisfaction of having reached the assigned objectives, either in terms of play or organisation, and of achieving a notable degree of collaboration between all organisations of public service in the Meuse basin. Once again, they should all get the credit of this success.

Transboundary Water Management in a Changing Climate – Dewals & Fournier (Eds)
© 2013 Taylor & Francis Group, London, ISBN 978-1-138-00039-1

Active flood management in Alpine catchment areas equipped with storage hydropower schemes

M. Bieri & A.J. Schleiss
Ecole Polytechnique Fédérale de Lausanne (EPFL), Laboratory of Hydraulic Constructions (LCH), Lausanne, Switzerland

F. Jordan
e-dric.ch Ingénieurs Conseils, Le Mont-sur-Lausanne, Switzerland

ABSTRACT: The simulation of runoff in Alpine catchment areas is essential for the optimal operation of high-head storage hydropower plants (HPP) under normal flow conditions, but also in case of flood events. A semi-distributed conceptual numerical approach has been developed, combining hydrological modelling and operation of hydraulic works. *Routing System* allowed runoff generation, simulation of the operating mode of complex HPP and the impact on the downstream river network for different scenarios. An overview on the hydrological modelling approach and its application for the upper Aare River catchment in Switzerland, where about half of the area is operated by a complex hydropower scheme, is given. The retention effect of the reservoirs and their management, including preventive turbine operations, on flood routing in the Aare River upstream of Lake Brienz is presented for the 2005 historical flood event.

1 INTRODUCTION

Flood events are highly damaging natural disasters. In mountainous areas, storage hydropower plants (HPP) strongly influence the flow regime of the downstream river system, depending on the drained area, the water storage capacities and their location within the basin. The regulation of hydraulic structures is therefore crucial in order to simulate runoff in catchment areas, e.g. in the framework of flood forecasting for emergency planning or long- as well as short-term inflow estimation for optimized reservoir operations. Most of the approaches lack the simulation of complex high-head hydropower schemes with large reservoirs for seasonal storage and low turbine capacities for peak-load, such as present in Alpine countries. Thus, the semi-distributed conceptual modelling tool *Routing System* was developed, simulating the hydrological processes as well as the operation of hydraulic structures. It was initially applied to the Rhône River basin upstream of Lake Geneva in Switzerland in the framework of the MINERVE project (Jordan et al., 2012). A decision-support system was implemented using the hydrological model and weather forecasts for real-time optimization of the preventive turbine and gate operations of the ten major hydropower schemes.

In the present study, the developed approach was applied to the upper Aare River catchment in Switzerland (Fig. 3), which is operated by a complex storage hydropower scheme. During the flood event of August 2005, the Aare River inundated the whole valley between Meiringen and Brienzwiler. The peak flow of $444 \, \text{m}^3/\text{s}$ was the highest measured discharge in Brienzwiler since 1905, corresponding statistically to a return period of about 100 years. Despite a levee break and flooding of agricultural land, major damages were avoided. In the context of a post-assessment of the event by the authorities and because half of the river basin is used for hydropower generation, the influence of the HPPs on flood had to be estimated by systematic simulations. This article gives some selected information about active flood in Alpine catchment areas equipped with storage

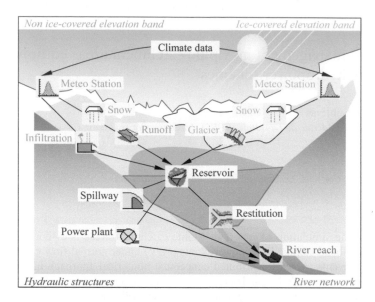

Figure 1. Function sketch of *Routing System* for high-mountainous catchment areas, indicating the meteorological data input, the hydrological elements in the ice- and non ice-covered elevation bands, the hydraulic elements of the HPP as well as the downstream river network for streamflow.

hydropower schemes. Extended explanation is given in various publications (Bieri et al., 2010, 2011; Bieri & Schleiss, 2012).

2 METHODS

2.1 *Simulation*

Routing System is a hydrological-hydraulic modelling tool for simulation of operated high-mountainous catchment areas. The semi-distributed conceptual hydrological modelling approach contains the reservoir-based precipitation-runoff transformation model GSM-SOCONT (Schaefli et al., 2005). Spatial precipitation and temperature distributions are taken into account for simulating the dominant hydrological processes, as glacier melt, snowpack constitution and melt, soil infiltration and runoff, as indicated in the upper part of Figure 1. Each basin is divided into ice- and non ice-coverd elevation bands, segregating rain and snow from corresponding precipitation and temperature. *Routing System* is able to take into account also hydraulic structures, such as water intakes, reservoirs, turbines, pumps, gates and their regulation, as indicated in the lower part of Figure 1. Thus, it allows simulation of the operation mode of HPPs and its impact on the downstream river system for different scenarios (Bieri & Schleiss, 2012).

2.2 *Plant operation*

An autonomous tool for optimal HPP operation has been developed (called *OptiProd*). Figure 2 shows an overall sketch of its operation mode. As a general rule, turbine operations should be performed during peak price hours to generate maximum revenue. Thus, the priority driving parameter is the demand by the electricity market. For this study, electricity prices were real spot market values from the European Energy Exchange (EEX). For the given forecast time horizon, the hours when the electricity price of the market is higher than the defined cost price, are set. A target level

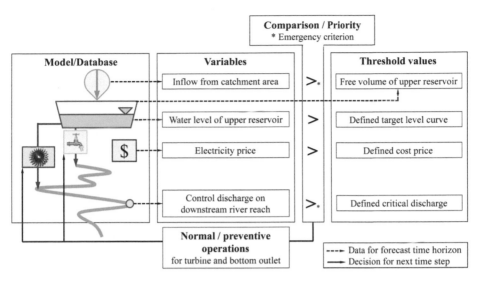

Figure 2. Flow chart of HPP operation for normal and preventive operations (*OptiProd*), containing time-dependent and pre-defined values.

curve, defining the annual filling of the reservoir, should guarantee the seasonal water transfer from summer to the economically interesting winter hours. The algorithm follows this curve with a pre-defined tolerance.

Besides normal operation, turbines and bottom outlets are used in case of emergency. If predicted inflow to a reservoir is higher than its free storage volume and thus overflow would happen, preventive emergency operations start, as long as the downstream located reservoir is not full or peak flow in the downstream river reach is lower than a pre-defined threshold value. A 0 h forecast horizon corresponds to instantaneous turbine operations, whereas 24 h or 48 h simulate in a first run the prediction data, which is then used as a decision support for the second run.

3 CASE STUDY

The upper Aare River springs in the glaciers of Unteraar and Oberaar at the altitude of 2000 m a.s.l. and flows nowadays through several artificial reservoirs (Oberaar, Grimsel, Räterichsboden), in which the main part of the water is temporally retained to be operated (Fig. 3). In Innertkirchen the water is given back to the Aare River immediately downstream the confluence with the Gadmerwasser, the river draining the eastern part of the catchment area. After the Aare Gorge the Aare River reaches the main valley of Meiringen and enters Lake Brienz at Brienzwiler. The surface of the upper Aare River basin is 554 km², where 21% was ice-covered in 2003. The hydrologic regime is glacial. The average annual discharge is 35 m³/s.

At the end of the 19th century, the area of the Grimsel and Sustenpass was recognized as particularly appropriate for hydropower production. Heavy rainfalls, large retention areas, solid granitic bedrock as well as large differences in altitudes over short horizontal distances provide optimal conditions for storage hydropower. The first concrete dams were built by the *Kraftwerke Oberhasli AG* (KWO) between 1925 and 1932. Since then, a complex scheme with nine power plants and eight reservoirs has been constructed (Fig. 3). The largest reservoirs are the lakes Oberaar (57 Mm³), Grimsel (94 Mm³), Gelmer (13 Mm³) and Räterichsboden (25 Mm³). The scheme comprises a large number of power houses and flood evacuation facilities. Table 1 shows the relevant elements and their admitted capacities for preventive operation.

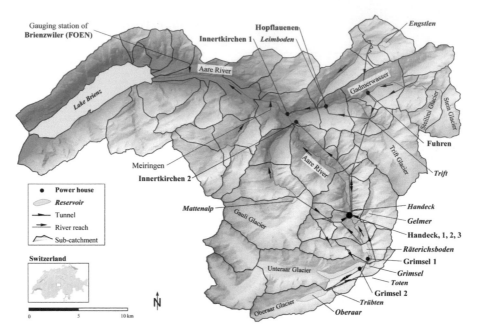

Figure 3. Map of the upper Aare River catchment with today's KWO hydropower scheme (reservoirs, power houses and tunnels), the sub-catchments areas and the river network with its location in Switzerland.

Table 1. Power houses and bottom outlets of KWO with their capacities for preventive operation.

Power house	From	To	Capacity [m^3/s]
Grimsel 1, Turbine 1	Oberaar	Räterichsboden	8
Grimsel 1, Turbine 2	Grimsel	Räterichsboden	20
Grimsel 2	Oberaar	Grimsel	93
Grimsel 3	Oberaar	Räterichsboden	–
Handeck 1	Gelmer	Handeck	18
Handeck 2	Räterichsboden	Handeck	32
Handeck 3	Räterichsboden	Handeck	14
Innertkirchen 1	Handeck	Aare River	39

Bottom outlet	From	To	Capacity [m^3/s]
Oberaar	Oberaar	Grimsel	26
Grimsel	Grimsel	Räterichsboden	28
Räterichsboden	Räterichsboden	Handeck	35
Gelmer	Gelmer	Handeck	20

4 MODELLING

4.1 Data sources

The meteorological input data was available from the Federal Office of Meteorology and Climatology. On the one hand, temperature and rainfall data were collected every ten minutes by an automatic monitoring network (ANETZ) all over Switzerland. On the other hand, a large number

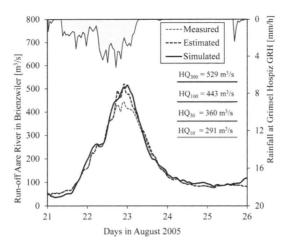

Figure 4. Calibration with 2005 flood event (NSE = 0.98, r_{vol} = 1.03, r_{peak} = 0.99).

of gauging stations (NIME) measure the daily rainfall. Five stations of the first type and nine of the second are used as input data points in and around the Hasliaare catchment. The discharge is measured every ten minutes on the Aare River in Brienzwiler by the Federal Office of Environment (FOEN) (Fig. 3).

The KWO made available the hydraulic characteristics of the hydropower scheme, operation rules and historical data from the last 30 years of exploitation. The datasets allow calculation of the inflow of the ten sub-catchments operated by KWO. Electricity prices were real spot market values from the European Energy Exchange (EEX).

4.2 Calibration and validation

The catchment area of the Aare River upstream of Lake Brienz was modelled in its configuration of 2003. Related to hydrological and HPP constraints, the river basin was divided in 43 sub-catchments, which were split in 96 ice-covered and 243 non ice-covered elevation bands of 300 m. For each band, precipitation, temperature and potential evapotranspiration were interpolated from the 14 meteorological stations, if located in the influence zone of 20 km search radius. The basic hydrological formulas as well as the calibration process are explained in Bieri & Schleiss (2012).

The model was pre-calibrated over a 12 month period for the ten sub-catchments operated by KWO (Fig. 4) as well as the natural catchment area upstream of the gauging station of Brienzwiler. In a second step, the model was calibrated by the extreme flood event of August 2005 and validated for the flood of August 1987, applying the same HPP operations as in the past. The measured peak flow of the upper Aare River in 2005 was 444 m³/s (called measured discharge) (Fig. 4). Because of lateral flooding, the entire discharge could not be measured at the gauging station. A post-analysis of the event, however, allowed an estimation of the real peak and a reconstruction of the hydrograph (called estimated discharge). Flooding is not simulated with *Routing System*. For this reason, the model was calibrated using the estimated hydrograph with a peak discharge of 520 m³/s. The time step was 600 s. Data acquisition for all elements in the catchment area was done with hourly mean values for the total simulation period. The results were compared to the inflow from the sub-catchments, to the observed outflow in Brienzwiler and to the peak flow estimations in terms of Nash and Sutcliffe efficiency NSE, water volume ratio r_{vol} and peak flow ratio r_{peak}.

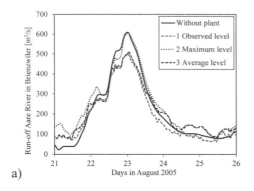
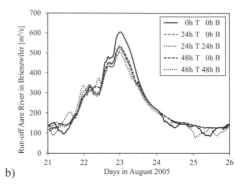

a)   Days in August 2005   b)   Days in August 2005

Figure 5. Hydrographs of the 2005 flood event for different initial reservoir levels (a) and for different prediction times for the maximum initial reservoir levels (b) (T = turbine; B = bottom outlet).

4.3 Scenarios

Preventive operation consists of lowering the reservoir levels by turbine operations or water release by bottom outlets before the flood peak is achieved. By avoiding outflow from the reservoirs during maximal flow period, peak discharge in the downstream river can be reduced. At the end of a flood period, the reservoirs should ideally be filled. For defining the potential of flood retention of today's hydropower scheme by preventive operation, several scenarios had to be tested:

- Three scenarios of filling degrees of the four main reservoirs Oberaar, Grimsel, Räterichsboden and Gelmer were defined, simulated and compared:
  - Scenario 1 corresponds to the measured or estimated levels of 2005 and 1987.
  - Scenario 2 is a worst case scenario assuming maximum reservoir filling on 18th of August 2005. This quite hypothetic case is the upper limit of the sensitivity analysis.
  - Scenario 3 presents average levels in August, calculated over the last 10 years, corresponding to the most likely filling degrees between 70 and 90%.
- Assuming, that flow can be adequately forecasted for a certain time horizon, preventive operation can be optimized. For the given cases, different combinations of prediction times of 24 or 48 hours for the turbine and bottom outlet operation were tested. For a time horizon of X h for the turbines (T) and Y h for the bottom outlets (B).

5 RESULTS AND DISCUSSION

The initial level in the main reservoirs is an important parameter, which influences directly the outflow of the system. Maximum reservoir levels without preventive operation produces 30% (Fig. 5a) higher flows in the Aare River than the average August reservoir levels, which generate a discharge of 500 m³/s for both events, due to flood routing in the reservoirs. The accumulated volumes are also higher, because of important flood release by the spillways, especially during increasing flows. Both downstream hydropower plants Innertkirchen 1 and 2 are operating on their maximum capacity. This preventive emptying of reservoirs leads to lower discharge in the Aare River. The peak flow of the 2005 flood could be considerably reduced from 605 m³/s, without preventive operation, to values between 535 m³/s for 24 h T and 0h B and 500 m³/s for 48 h T and 48 h B (Fig. 5b). Similar simulations have been done for the extended scheme of KWO and published in Bieri et al. (2011).

6 CONCLUSIONS

The presented model is robust and gives satisfying results for the simulation of the observed flood events. By taking into account the influence of a hydropower scheme, different scenarios could be analysed. Constraints and initial conditions as well as input data can be adapted and their effects evaluated.

Preventive operation for lowering the reservoir levels could reduce the flood peak in the Aare River. Higher flood peak reduction is achieved by using not only the turbines but also the bottom outlets of the dams. Extended schemes with higher operation capacities and storage volume could increase the retention effect (Bieri et al., 2011).

An active flood management would require a decision-making strategy, including legal and economic threshold values and constraints for preventive measures. The developed modelling approach proves the potential of passive and active flood retention of storage hydropower schemes.

REFERENCES

Bieri, M., Schleiss, A.J. & Fankhauser, A. 2010. Modelling and simulation of floods in alpine catchments equipped with complex hydropower schemes. In A. Dittrich et al. (eds.), *River Flow*; *Proc. intern. symp., Braunschweig, 8–10 September 2010*. Karlsruhe: Bundesanstalt für Wasserbau.

Bieri, M., Schleiss, A.J., Jordan, F., Fankhauser, A.U. & Ursin, M.H. 2011. Flood retention in alpine catchments equipped with complex hydropower schemes – A case study of the upper Aare catchment in Switzerland. In: Schleiss & Boes (eds.), *79th ICOLD Annual Meeting, Lucerne, 1 June 2011*. Rotterdam: Balkema.

Bieri, M. & Schleiss, A.J. 2012. Analysis of flood-reduction capacity of hydropower schemes in an Alpine catchment area by semi-distributed conceptual modelling. In *Journal of Flood Risk Management*: online.

Jordan, F., Boillat, J.-L. & Schleiss, A.J. 2012. Optimization of the flood protection effect of a hydropower multi-reservoir system. In *Int. Journal of River Basin Management* 10(1): 1–8.

Schaefli, B., Hingry, D., Niggli, M. & Musy, A. 2005. A conceptual glacio-hydrological model for high mountainous catchments. In *Hydrology and Earth System Sciences* 9: 95–109.

Transboundary Water Management in a Changing Climate – Dewals & Fournier (Eds)
© 2013 Taylor & Francis Group, London, ISBN 978-1-138-00039-1

Towards a roadmap to climate change adaptation in the Meuse river basin, with the focus on water quantity

I. Krueger, B.T. Ottow & O. de Keizer
Deltares, Delft, The Netherlands

H. Buiteveld
Rijkswaterstaat, Lelystad, The Netherlands

ABSTRACT: This paper presents the development process and building blocks of a first approach to a roadmap towards climate adaptation in the Meuse river basin. In order to prepare the way for a transnational roadmap, several interviews and two workshops with policy-makers and researchers were held. In these workshops a common vision was developed, and the challenges and necessary measures formulated and the desirable steps per actor were identified. This was done as part of the INTERREG project AMICE.

1   INTRODUCTION

Climate change is one of the main water management challenges in the Meuse river basin that have to be dealt with in the coming years. The AMICE project covered the complete range from setting up climate scenarios, assessing the effects on water quantity and identifying water management issues up to the necessary adaptation. This paper describes how the process of setting up an international adaptation strategy was initiated (Krueger at al., 2012). Arriving at an internationally coordinated adaptation strategy requires several steps. The role of the AMICE project was to initiate the process, in which decisions on the international adaptation are taken at the respective governmental levels. For this subject, the AMICE project focused on the way how this can be achieved.

Just as a road map for motorists shows the users where they start and where they want to arrive, so did the AMICE project intend to give an impression of what is needed to set out the pathway to climate adaptation in the Meuse Basin, by outlining the situation of departure, the possible desired outcome (where to go to, the vision), the possible obstacles on the way, and the choices that need to be made, in order to achieve climate adaptation.

This paper summarizes this process in four components; the current situation of the Meuse river basin with respect to climate change adaptation, exclusively based on the results produced in the AMICE project. Next, a joint vision on the future of the Meuse river basin in the year 2100 is presented which gives an impression on the desired outcome. Thereafter, the challenges which are faced on the way to reach the desired outcome are presented, and different steps are identified on national and trans-national level.

2   METHOD

2.1   *Stakeholder workshops*

The AMICE Partnership does not represent all stakeholders and decision-making levels in the Meuse river basin. In order to develop a roadmap which takes into account all on-going projects and policies, It was decided to open the discussions to a wider audience. Two workshops were held in which 4–5 policy-makers per country/region participated together with a number of researchers from the

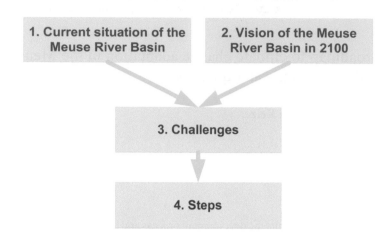

Figure 1. Overview of the components for a transnational roadmap for climate-adaptation in the Meuse river basin.

AMICE project. In preparation for the workshops, 9 interviews with policy-makers were held. In total, 28 policy-makers from national/federal ministries, state/province government, regional water management organizations and municipalities in France, the Walloon Region, Flanders Region, Germany and the Netherlands participated in the interviews and/or the workshops.

Based on the exploratory interviews, a starting document was prepared which offered background information and formed the basis of the discussion during the workshops.

In the first workshop, participants exchanged their visions of the desired situation of the Meuse river basin in the year 2100 and compiled a list of priority trans-national challenges from climate change in the Meuse river basin. Furthermore, a list of measures which could be implemented to prevent or mitigate negative impacts from climate change were identified. Along with the purpose of producing a vision, challenges and measures, the workshop also served to better acquaint the water managers and decision-makers from the different parts of the river basin with each other.

2.2 Component of the roadmap

Subsequent to this workshop, the AMICE project team convened and discussed the workshop results. Measures were analysed and different clusters of measures formed. The results of this analysis were presented to the policy-makers in the adapted starting document and during the second workshop which took place a month and a half after the first workshop. As the participants of the two workshops differed due to different availabilities, the second work-shop allowed also for input into the results of the first workshop. Thereafter, the steps which should be taken to enable the implementation of measures were identified on a (sub-)national, bilateral and transnational level in country groups. Summarizing, the process produced four components for a roadmap towards climate adaptation in the Meuse River Basin, as presented in Figure 1.

3 COMPONENTS ROAD MAP

3.1 Current situation of the Meuse river basin

The point of departure for the transnational roadmap for climate adaptation in the Meuse river basin is the current situation of the Meuse river basin with respect to climate change adaptation, as depicted by the results of the AMICE project. A more detailed overview of the results of the project can be found at: www.amice-project.eu.

It should be kept in mind that the exact implications of climate change are and will remain uncertain. AMICE helps to get a better picture of what the trends are, but cannot predict which scenario is actually going to happen in the Meuse river basin, a problem which is also experienced in other European river basins, such as the Rhine and the Danube. AMICE made climate projections for the medium (2021–2050) and far (2071–2100) future. Two extreme scenarios are built by the AMICE partners that represent the reliable envelope of possible futures. It is important to realize that the scope of AMICE is water quantity. Together, the AMICE partners have agreed on the following hydrological scenarios for the whole Meuse area:

- Flood discharges: an increase in Q100 (centennial hourly flood peak) of +15% for 2021–2050 and +30% for 2071–2100
- Low-flow discharges: a decrease in MAM7 (Mean Annual Minimum 7-days discharge values) of −10% for 2021–2050 and −40% for 2071–2100.

In the scope of AMICE, flood maps along the whole course of river Meuse were developed, accounting for the hydrological impact of climate change. The flood maps and calculations show a significantly higher impact of climate change on water depth in the central part of the basin (max. 130 cm over the centennial flood), compared to the upper and lower parts (max. 70 cm over the centennial flood). This is due to the morphology of the valleys, which are narrower in the central part, as compared to upstream and downstream floodplains. Based on these flood maps, potential economical damage caused by the extra flooded area was calculated. With regard to monetary damages, the major cities along the river are the most vulnerable to future floods. The Meuse basin is largely covered by forest and agricultural land, however human settlements account for the major part of the total flood damage.

3.2 Vision of the Meuse river basin in 2100

The joint vision of the Meuse Basin in 2100 is based on a brainstorm exercise in the first workshop with stakeholders on the long term goals for the Meuse River basin. In this brainstorm exercise, participants were asked to express their dream vision of the Meuse basin in the year 2100. In this way they formulated a joint wish and hope for the future of the Meuse river basin, which outlines the direction, which future development in the Meuse river basin should take. At the same time, such a joint stakeholder vision helps to unravel underlying contradictions which show the different future desires of different stakeholders. By this, the need for measures which can help to overcome what we currently see as a contradiction becomes clear.

It is clear that one brainstorm is not enough to produce a coherent vision, because of addition that we made afterwards end also contradictions in the vision. An addition to the vision which was made after the workshops was that the vision should explicitly name flexibility in the approach, resilience as aim and overall sustainability of actions/measures. A remarkable characteristic of the vision is the apparent contradictions between different ideas of the future, such as more natural river characteristics together with more shipping. The visioning exercise also pointed at the existing knowledge gaps with respect to the development of population dynamics in the river basin, and the future socio-economic changes to be expected. Figure 2 presents an image of the desired future for the Meuse river basin (artist impression). The vision of the Meuse shows a river basin without national and sub-national borders. This is to be understood as a metaphor of the desire for a stronger international cooperation and coordination of water management in the river basin.

The question arises, how these inherent contradictions can be overcome in the Meuse river basin. Are there measures that can combine seemingly non combinable sectoral needs? Can mitigation take place between different desires, or can the same outcome be achieved with different measures than those foreseen in the vision? How to organize joint, internationally coordinated decision-making can address these contradictions and produce adequate trade-offs, with the final aim to promote sustainable river basin development? The visioning exercise also pointed at the existing knowledge gaps with respect to the development of population dynamics in the river basin, and the future socio-economic changes to be expected.

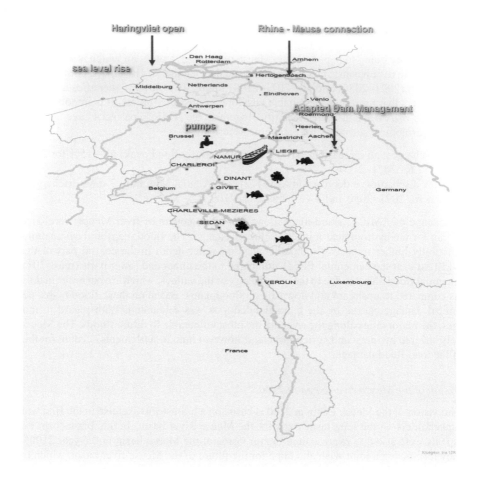

Figure 2. Artist impression of the vision on the possible future of the Meuse River Basin in 2100.

3.3 *Challenges*

Based on the vision, different challenges which the Meuse river basin is facing, and which prevent it from reaching this vision were identified in the first stakeholder workshop. Stakeholders classified most of the challenges as transnational in nature, rather than regional. The majority of those challenges can be solved regionally, and many of them exist in all countries. There was a general consensus that most challenges are urgent, and implementation has to be started as soon as possible. The identified challenges were clustered as presented in Table 1.

Remarkably, in the interviews, which had been conducted previous to the workshop, a strong emphasis was put on the challenge that cooperation and coordination are posing to the river basin. Water shortage and water quantity were also mentioned as an important issue for the whole river basin. However the challenges are not distributed homogeneously over the river basin. Different countries experience different challenges, and the degree to which the different regions/countries perceive a challenge as urgent strongly varies. In and after the second workshop, several challenges were identified that were not included in the challenge clusters in Table 1. These additional challenges are listed in Table 2.

For the above mentioned challenges, measures were identified which could help mitigate or relieve them. In interviews, stakeholders pointed at possible criteria, which good measures would need to meet, in order to be accepted (Fig. 3).

Table 1. Overview of challenges identified during the first workshop.

Challenge cluster	Challenge
Land-use	Spatial planning/Better land use planning Floods/Damage caused by floods/Multifunctional zones (agricult. Supply & water retention)/Lack of land, e.g. (for reservoirs, etc.) Change agriculture (EU)/Room for water –> restore water system (river-valley – groundwater)
Awareness	Public awareness (water use, water problem, climate change) more efficient on long term/Awareness & change of life/Climate Sceptic/Willing to change (behavior) and take actions/Sociology, basin culture/Promote Meuse Basin as example/Lack of money
Water Quality & Ecology	Regional differences, it depends on river or river stretch/Not for all substances there are standards (e.g. emerging substances, pharmaceutics, etc.)/Not all stan dards are met/Adaptation of agricultural practices/More water efficient processes for industry
Improve knowledge	Missing knowledge: lots of open questions/Impact of substances/Prediction of future trends/Impacts of measures on hydro-morphology/Cost-effectiveness, Economic analysis of water services/Relationship between Agriculture and infiltration, Agriculture and erosion, Agriculture and suspension/sediments

Table 2. Overview of additional challenges.

Challenges that could not be included in a Cluster	Lack of money in communities Increase of water demand Yet unforeseen challenges in other sectors than water Different impacts in different regions
Challenges identified during the definition of Measures	Heat-island effect in cities Biodiversity Climate change mitigation
Challenges identified by the AMICE Partners in earlier brainstorming sessions	Uncertainty in the climate projections and in the socio-economical scenarios: integrate uncertainty in planning and designing, update plans on a regular basis Awareness and Involvement of politicians Early warning systems (esp. for flash floods)

Figure 3. 'Word cloud' of the desired characteristics of good solutions, as identified by interviewees from the different countries/regions. Size of the words correlates with the number of interviewees who named the issue (total number of interviews: 9).

3.4 *Towards a roadmap*

In four national groups, the participants of the second workshop identified steps that should be taken regionally/nationally, bilaterally between nations, and multilaterally between nations, to achieve an effective climate-change adaptation program for the Meuse River Basin. A comparison of the lists of steps identified per country shows that there are differences, but also commonalities between countries:

- At national levels the organizations responsible for climate change adaptation are in place, but implementation is not yet effective and/or concrete. There is a need for action, concrete steps need to be identified per country.
- At bilateral levels there is a strong wish for better communication, sharing of information, knowing each other and each other's institutional set-up, which is broader and not only related to climate adaptation.
- At transnational level there is a wish for better coordination, more incentives and a harmonization of policies and planning where possible and needed.

The observed differences can mostly be attributed to the different local contexts: Depending on their geographic position in the river basin (upstream/downstream), their water governance structure and the functioning of their part of the water system, countries are faced with different challenges with respect to climate change. For the same reasons, also the degree of urgency associated with the challenges may vary between countries. On the other hand, several common challenges can be observed, which offer opportunities for future cooperation. Another example, apart from the above mentioned is the need to clarify the urgency of climate change adaptation and the necessity of actions to be undertaken, or the need for more insight and knowledge on the Meuse river basin. These needs were not only identified on the national levels, but also on bilateral and international level.

4 CONCLUSIONS

Based on the interviews and the workshops, we come to the following conclusions.

Decision-makers in the Meuse river basin share a common vision with respect to how they imagine the river basin to look like in the year 2100. The inherent contradictions in this vision were recognized and discussed in workshops, and are also reflected in the resulting list of challenges. These challenges can be grouped in different clusters, of which 'water quantity' (too much and too little water) and 'coordination and cooperation' were most prominent. The need for better international coordination and cooperation which was identified in the challenges also returned when identifying steps on trans-national level. Also on bilateral level, there was a wish for better communication and a better acquaintance between organisations.

The steps identified for national and bi-lateral level showed commonalities and differences, depending on the geographical location in the river basin. For the further-development of the roadmap to climate change adaptation, we recommend to take the following steps:

i. AMICE should make its results and the importance thereof clear and operational.
ii. The respective managing authorities should organize or use existing national workshops and/or working groups to discuss the results of AMICE and identify the appropriate measures, both national, bilateral and transnational
iii. Transnational bodies must identify common coordination actions that assist the above.

REFERENCE

Krueger, I., Ottow B. & Fournier M. (2012). *AMICE: Towards a transnational roadmap for climate adaptation in the Meuse river basin*. Report Amice.

Transboundary Water Management in a Changing Climate – Dewals & Fournier (Eds)
© 2013 Taylor & Francis Group, London, ISBN 978-1-138-00039-1

The Programme Interreg IV B North-West Europe

L. Tuulik
Joint Technical Secretariat Interreg IV B NWE, Lille, France

ABSTRACT: INTERREG IVB North West Europe (NWE) Programme is a financial instrument of the European Union's Cohesion Policy.

Over the seven years (2007–2013), the programme invests €355 million from the European Regional Development Fund (ERDF) into the economic, environmental, social and territorial future of North West Europe (NWE). The fund is used to co-finance projects that maximize the diversity of NWE's territorial assets by tackling common challenges through transnational cooperation. To this end, the Programme seeks organisations that are resolute in their ambition to contribute to a cohesive and sustainable territorial development of North West Europe.

1 PROGRAMME PRIORITIES

1.1 *Transnational cooperation*

Transnational cooperation is the core of the INTERREG IVB Programme. It allows countries to work together on mutually beneficial projects to tackle issues that go beyond national borders. It produces transferable working models, and speeds up the process of innovation through the sharing of knowledge and development costs. The collective benefits of such collaboration are invaluable; participating organisations acquire new skills, initiate effective working methods and increase their connections to European network.

1.2 *Transnational challenges*

A transnational challenge, by definition, can only be tackled effectively at local, regional or national level through co-operation. That is to say, the scale and consequences of any given social, economic or environmental challenge should determine the scale of the intervention. North West Europe suffers from common environmental, social and economic pressures that are neither confined by national governments nor by administrative boundaries. The NWE Programme helps project promoters to transform such transnational challenges into opportunities for change.

The results of previous projects are examples of this ambition: some prevented river flooding by widening the riverbed; other projects used wetlands to create resources for adapted agriculture and forestry; several others tackled social issues by redeveloping urban brownfields to foster opportunities for small-scale economic development. To succeed, these challenges often necessitate multilateral intervention. With an aim to facilitate such intervention, the European Union funds projects in the field of territorial development through the INTERREG IVB NWE Programme.

1.3 *Thematic priorities*

1.3.1 *Capitalizing on innovation*
The NWE region is the economic powerhouse of Europe: it is endowed with a wealth of territorial assets including a highly skilled workforce and world class universities. To extract the economic

the value of such attributes, projects under this priority should aim to strengthen the economic competitiveness of NWE in response to the Lisbon agenda for growth and jobs.

Projects should aim to produce transnational partnerships which can enhance the region's capacity to innovate and facilitate the development of knowledge-based activities. Preference will be given to projects which can develop cross-sectoral synergies and facilitate the creation, demonstration and above all, the application of knowledge.

Project activities should be clearly linked to territorial development and not solely focus on networking within the specific scientific/business sector, for which, the interregional cooperation strand is more appropriate. The Programme will not support research and development or academic networking which is not linked to actions or demonstration projects.

1.3.2 *Managing natural resources and risks*

The sustainable management of natural resources and risks is of utmost importance for the NWE area. This priority calls for intervention based on a broad range of activities which attempt to minimize and prevent the pollution of land, water and air. In the case of coastal, marine and river flooding, preference will be given to projects which transfer applicable knowledge and develop innovative responses across the whole of NWE area.

Preference will be given to cross-sectoral projects addressing obstacles in legislative systems and those which focus on improving the lack of integration in institutional and governmental structures. Projects limited to data collection and management, or local/regional activities and flood defence investments that are not relevant to the wider transnational cooperation area will not be supported.

AMICE has been approved under this Priority 2.

1.3.3 *Improving connectivity*

North West Europe is, in general, characterised by a high level of accessibility. Regional imbalances, however, continue to persist: the core area suffers from congestion and the more peripheral and rural areas endure low levels of accessibility.

The focus of the Programme is to overcome such challenges, not only through demand management and a more efficient and sustainable use of existing capacity, but through the provision of new small-scale infrastructure.

Projects which improve the transnational coordination of transport and ICT-related solutions will be well received. Similarly, the Programme will support projects which seek to strengthen the political and institutional framework for enhancing the quality and performance of such infrastructure and services.

1.3.4 *Strengthening communities*

This priority supports transnational actions that facilitate economic and social cohesion within and between cities, towns and rural communities. Projects should aim to enhance the potential of regional assets aim to improve the attractiveness of environments, examine the potential for energy efficiencies in the construction and use of buildings and find solutions to the impacts of demographic change and migration.

This priority, therefore, seeks actions and model solutions for adapting policies to achieve a better balance in the settlement structure and avoid further polarisation and depopulation tendencies. Projects which concentrate on local actions with no transnational synergy effects, and which do not provide wider territorial development benefits for the transnational area will not be supported.

2 MORE INFORMATIONS

http://www.nweurope.eu

Liina Tuulik has been working as a project officer in different Interreg programmes since 2006. She joined the NWE programme in the beginning of 2012 and her tasks include the project development and guidance as well as monitoring of the running projects.

Transboundary Water Management in a Changing Climate – Dewals & Fournier (Eds)
© 2013 Taylor & Francis Group, London, ISBN 978-1-138-00039-1

Deltares activities in the Meuse river basin

F.C. Sperna Weiland, M. van Dijk, O. de Keizer & A.H. Weerts
Deltares, Delft, The Netherlands

ABSTRACT: Deltares is a research institute for water and subsurface issues operating in deltas, coastal areas and river basins worldwide. Its projects are related to technological issues, natural processes, spatial planning and administrative and legal processes. In this proceeding we present several Deltares activities in the Meuse basin; 1) operational flood forecasting, 2) design discharge estimation and 3) climate impact assessments.

1 DELTARES

Deltares is a research institute for water and subsurface issues with its base in the Netherlands. Throughout the world, Deltares enables sustainable living in deltas, coastal areas and river basins. Deltares advises governments and the private sector, and uses its expertise to make sound and independent assessments. Deltares projects involve not only technological issues, but also natural processes, spatial planning and administrative and legal processes. Deltares is concerned with areas where economic development and population pressure are high, where space and natural resources, both above and below the surface, have to be used and protected in multi-functional and intensive ways. Deltares employees have been working in the river basins of the Meuse and Rhine for decades. Below we will give a short introduction of the variety of Deltares activities in the Meuse basin.

2 FLOOD FORECASTING – DELFT-FEWS

Deltares developed a waterlevel and flood forecasting system for the Dutch Waterdienst and the Water Management Centre Netherlands (WMCN). The forecasting system is configured in the real time software infrastructure DELFT-FEWS for operational water management and forecasting (Werner et al., 2012) that has been implemented in over 20 coutries amongst which the UK, USA and Australia. The system integrates hydrological and hydraulic models (HBV, SOBEK-1D) with software for the import, validation, interpolation and presentation of data. The most relevant data and model results are stored in a database, the core of the system. The system continuously collects measured and forecasted waterlevels for about 100 locations along the river Rhine and Meuse and meteorological data from the KNMI, Deutscher Wetterdienst (DWD) and MétéoFrance. Weather forecast data from the KNMI, DWD and ensemble forecasts from European Centre for Medium-Range Weather Forecasting (ECMWF) are used as input for the (probabilistic) hydrological forecasts. Based on the forecasted waterlevels Dutch water managers can decide whether further action is required. Currently, the FEWS system is being improved in coopcration with Wageningen University and the TU Delft (Rakovec et al., 2012a, 2012b).

3 DESIGN DISCHARGE ESTIMATION

In the Netherlands the flood protection situation is evaluated every five years. This also involves the evaluation of the design water levels for dike dimensioning along the Meuse. Originally this was done with traditional methods using frequency analysis of extreme discharges. Since 1996 the Dutch Waterdienst, the Royal Dutch Meteorological Institute (KNMI) and Deltares have been working on a new method for the estimation of design discharge for dike safety. This method involves a stochastic weather generator, that generates long synthetic rainfall and temperature time-series, and a suite of hydrologic and hydraulic models that transform these time-series into discharges. The instrument is called the Generator of Rainfall and Extreme Discharges – GRADE (Wit and Buishand, 2007). With this instrument long discharge records can be simulated from which flood frequency curves can be derived that should be more reliable for high return values (Kramer et al., 2008). The methodology can also be applied to assess the impact of climate change on future flood frequencies. After a large number of uncertainty analysis, involving meteorological, sampling algorithm and parameter uncertainty, the instrument will now soon become part of the National Dutch Evaluation Instrument for Flood Safety.

4   CLIMATE IMPACT ASSESSMENT

Deltares was one of the partners of the AMICE project. Here the Deltares activities focused on the hydrological and hydraulic climate modeling of the Meuse river. Currently Deltares and the Dutch Waterdienst are initiating a follow up of the AMICE project – the Vue de Meuse project. The aim of this project is to develop consistent hydrological trans-boundary climate projections for the river Meuse. These projections will provide information on future changes in flood frequencies, droughts and low flow occurrence. The study will be based on a large number of climate models and scenarios (Van Pelt et al., 2012), and ideally a number of hydrological models, in order to assess the uncertainty of the climate projections. Within this project Deltares and the Dutch Waterdienst wish to actively co-operate with institutes and universities from the other Meuse countries in order to enhance the international acceptance of the project results.

REFERENCES

Kramer, N., Beckers, J. & Weerts, A. 2008. Generator of Rainfall and Discharge Extremes, Part D&E. Deltares report.
Rakovec, O., Weerts, A.H., Hazenberg, P., Torfs, P.J.J.F. & Uijlenhoet, R. 2012. State updating of a distributed hydrological model with Ensemble Kalman Filtering: Effects of updating frequency and observation network density on forecast accuracy. *Hydrol. Earth Syst. Sci.*, 16, 3435–3449, doi:10.5194/hess-16-3435-2012.
Rakovec, O., Hazenberg, P., Torfs, P.J.J.F., Weerts, A.H. & Uijlenhoet, R. Generating spatial precipitation ensembles: impact of temporal correlation structure. *Hydrol. Earth Syst. Sci.*, 16, 3419–3434, doi:10.5194/hess-16-3419-2012.
Van Pelt, S.C., Beersma, J.J., Buishand, T.A., van den Hurk, B.J.J.M. & Kabat, P. 2012. Future changes in extreme precipitation in the Rhine basin based on global and regional climate model simulations. *Hydrol. Earth Syst. Sci. Discuss.*, 9, 6533–6568, doi:10.5194/hessd-9-6533-2012.
Werner, M., Schellekens, J., Gijsbers, P., van Dijk, M., van den Akker, O. & Heynert, K. 2012. The Delft-FEWS flow forecasting system. *Environmental Modelling & Software*. 1–13, doi: 10.1016/j.envsoft.2012.07.010.
Wit, M. de & Buishand, T.A. 2007. Generator of Rainfall And Discharge Extremes (GRADE) for the Rhine and Meuse basins. *RWS RIZA, Report 2007.027, ISBN 9789036914062/KNMI publication: PUBL-218, ISBN 9789036914062*, 7/2007.

Knowledge of the river Meuse

H. Nacken
Academic and Research Department Engineering Hydrology, RWTH Aachen University, Germany

ABSTRACT: Prior to AMICE studies had already been undertaken relating to future climate change, summarised in 'The impacts of climate change on the discharges of the river Meuse' (2005) by the International Meuse Commission (IMC). Adaptation strategies were undertaken in the different countries of the Meuse basin but were barely concerted at the transnational level. Therefore one conclusion was that there is a need to agree on common climate change scenarios between the member states. Within AMICE such scenarios were considered.

1   AMICE INCREASES KNOWLEDGE

The Meuse is one of the largest rivers in NEW, with a catchment basin incorporating 5 member states, and situated in a densely populated area. Its discharge fluctuates considerably with seasons: it reached $3100\,\mathrm{m^3/s}$ in winter 1993 at the Dutch/Walloon border and is only $10\text{--}40\,\mathrm{m^3/s}$ in summers. Classed as a rain-fed river, it has no glacier and a little groundwater storage capacity to buffer precipitation. A direct link between climate evolutions and changes in high and low-flows exists, putting the assets of the basin, including major infrastructures, industries, priceless historical and ecological heritage, at risk.

As recommended by the IMC with the AMICE project basin-wide scenarios on climate change and discharges were developed. In order to characterize the possible evolution of the future climate-induced changes of the high and low-flows for the two time periods 2021–2050 and 2071–2100 a transnational wet and a dry and a national wet and a dry scenario were considered.

The AMICE partners agreed upon the most extreme hydrological scenarios, which were derived from the transnational wet and dry scenarios. They are as follows:

– An increase in Qhx_{100} values of $+15\%$ for 2021–2050 and $+30\%$ for 2071–2100
– A decrease in MAM_7 values of -10% for 2021–2050 and -40% for 2071–2100.

These hydrological scenarios were used by the AMICE partners involved in hydraulic modelling. After developing a common methodology the first coordinated flow simulation of the Meuse from spring to mouth was conducted. In the central part of the basin the increase of the water depth of a similar change in flood discharge is found to be approximately twice as strong as in the upper and lower parts. The simulations emphasised how some reaches may be severely affected by climate change in terms of increase of flooded areas, such as the reaches between Andenne and Monsin in Wallonia.

The AMICE project has proved that transnational cooperation benefits all member states. Not only because the individual works were concerted, but also because of the shared knowledge and the enhanced cooperation between all actors.

Transboundary Water Management in a Changing Climate – Dewals & Fournier (Eds)
© 2013 Taylor & Francis Group, London, ISBN 978-1-138-00039-1

The CONHAZ project

C. Green
Flood Hazard Research Centre, London, UK

1 INTRODUCTION

CONHAZ was a EC FP7 project to review the state of the art in assessing the costs of respectively floods, droughts, avalanches and coastal storms. Developing cost assessment methodologies are inherently exercises in transcience: the purpose of the exercise being to help decision makers to be able to make 'better' decisions around flood risk management in the future. So, the guidelines had to be relevant and appropriate to the different decisions that have to be made. Therefore, the process of developing the guidelines was stakeholder led, an obvious stakeholder being the CIS Working Group on floods; other stakeholders including land use planners, environmental interests and the insurance industry.

We began by holding workshops in which the two questions we asked of the stakeholders were:

– what do you want to know?
– how do you want to know it?

It is difficult for people to identify what they want unless they can see what they can have. It is easier to see the limitations and defects of something than to define in the abstract what they would like. Therefore, a draft outline of the guidance was distributed prior to the meeting so the stakeholders could judge whether we had some understanding of the decisions they had to make, how they framed those decisions, and what is important. This opening workshop resulted in significant developments in the guidelines.

Similarly, the entire project ended with a combined workshop of stakeholders in which a stakeholder was asked to review each of the draft sets of guidelines for the different hazards and then the cross-cutting issues were discussed. Thus, the success of the project is to be judged by whether the stakeholders find the outputs useful to the challenges they face.

2 CONTEXT

What economists do and what the stakeholders want from economic analysis have conventionally been two different things. But the social utility of economics lies in the extent to which it helps us to make better collective choices. This was the basis upon which the guidelines were developed.

So, the starting points for framing the guidelines and as discussed with the stakeholders were:

– Decisions are now made through a process of stakeholder engagement; critically, they decide what is the best available course of action to adopt.
– Decisions are necessary when the available courses of action are mutually exclusive but there apparent advantages and disadvantages to each course of action. Thus, the two conditions that make a decision necessary are conflict and uncertainty; decisions are processes in which we seek to resolve the conflicts that make the decision necessary and to become confident that one option should be preferred to all others.

- In making that decision, they want to know what are the likely consequences of each possible course of action.
- Ideally, they would also like advice as to what is the best process to go about resolving the conflicts and becoming confident that one option should be adopted.
- The aim in societal decisions is now to promote well-being.
- We are now seeking to make the transition to a green economy and sustainable development; we are seeking to change what we do and how we do it.
- We have to make this transition in the face of change; the other changes that are occurring such as climate change and an ageing population.
- The problem with water is to adapt to and cope with the inherent variability of water availability; a flood is an extreme perturbation as part of this variability. But floods and droughts are simply the two extremes of this variability; it is the variability which we must confront rather than approaching floods or droughts in isolation.
- Hence, change is at the centre of the problem: we want to make change, we want to learn and innovate so as to make better decisions where decisions always play out in the future, and we have to cope with change. The one constant is change. Change related concepts such as learning, innovation, adaptive management and resilience have to be at the centre of what we do and how we do it.
- The variability of the water environment is one aspect of water management and reflects the reality that catchments are dynamic systems of exchanges between land and the watercourses not only of water but also of soil and pollutants. As a system, they have to be managed as a system and local solutions risk simply moving the problem about.
- At the same time, we manage water largely in order to make the best use of land whilst the way in which we use land has dramatic impacts on the waterine environment. Hence, in flood risk management, the objective is not to reduce flood losses but to enhance the overall functioning of the catchment.

Economic analysis is a servant of this process: using the swan model, the swan gliding elegantly across the surface of the water, the economic analysis are the feet paddling frantically below the surface. In doing this, we used work around about the major outstanding unresolved theoretical problems in economics. Since it would have been inappropriate to have been prescriptive, at a number of points options were outlined. For example, there is a major division between those who consider that uncertainty can be handled through the use of probabilities and those who hold that uncertainty and probability are two entirely different things.

3   WHAT WE LEARNT

Having stressed the importance of learning, I will conclude by discussing what we learnt about the process of doing transience and the problems of assessing the ways in which shocks such as floods propagate through the economic, social and ecological systems and how those systems then recover from those shocks.

Using agro-hydrology to adapt to climate evolutions

A. Degré, C. Sohier, A. Bauwens & M. Grandry
Univ. Liège – Gembloux Agro-Bio Tech, Soil-Water Systems, Gembloux, Belgium

ABSTRACT: Natural phenomena such as floods, drought, erosion, nitrate leaching and plant growth are influenced by climate change. The Soil-Water Systems division of Gembloux Agro-Bio Tech aims at studying these phenomena; better understanding processes; modelling them in order to predict their change in the future and to assess their potential consequences. Then, we can propose strategies to adapt to these changes. As an agronomy faculty, we believe that adapting agriculture can play a major role in mitigating climate evolution effects at plot and catchment scales.

Soil-Water Systems is a division of the department of Environmental Sciences and Technologies of Gembloux Agro-Bio Tech (GxABT), the agronomy faculty of the University of Liège (B). Environmental management is, in fact, one of the two main research topics of the faculty, the other one being the valorisation of bioproducts, both linked with the production of bioresources. Management plans of arable, forested and natural lands proposed by GxABT will take into account global change (climate change, land use changes and urbanisation). Therefore, within the Soil-Water Systems division, projects focus on the link between hydrology and agriculture, taking account of climate change. Indeed, climate change disturbs most natural phenomena and it is important to quantify and mitigate these impacts.

As a partner of the AMICE project, we have contributed to the study of the hydrological effects of climate change on the Meuse catchment and developed a methodology to evaluate its impacts on the Mosan agriculture. Results have shown that both winter high flows and summer low flows could be exacerbated. Studying these extreme conditions is thus essential. Consequences on the agricultural sector can be both positive and negative, depending on the range of predicted changes and the adaptation capacity of agricultural systems. We have found that, in the Meuse catchment, yields would increase for wheat, barley and grasslands but decrease for corn. However, the variability in yields would rise in the future. The study has also revealed that the different crops would start growing and reach maturity earlier. Agricultural systems will therefore need to be adapted (e.g. adaptation of the cultural calendar by moving forward the seeding and harvesting dates) (Drogue et al., 2010, Bauwens et al., 2011).

The simulations for this study were performed with a physically-based model able to simulate the water-soil-plant continuum (derived from the EPIC model). The EPIC model has actually been adapted to the Walloon Region (EPICgrid) and has also been used to model nitrate concentration in the recharge water. EPICgrid has allowed us to evaluate diffuse nitrate pollution and the efficiency of present and further mitigation measures. Simulations results have shown that current measures to reduce the effect of diffuse pollution on water quality are not sufficient in some areas and that new actions are necessary. Scenarios including modifications of agricultural practices, such as a change in crop rotations or a decrease in the use of mineral fertilisers, have shown significant effects on water quality. Nevertheless, due to transfer time through the vadose zone (more than 20 years in some subregions), it was shown that an increase in groundwater nitrate concentrations will occur until at least 2030 for some regions before new agricultural practices can impact groundwater quality positively. Moreover, an increase in nitrate leaching in the future due to longer saturation periods in autumn and winter has been found under the wet scenario (Sohier et al., 2009, Sohier, 2011).

Yet, these results have to be taken with caution since future climate is uncertain. For example, the evolution of precipitation quantity is not known with certainty. Nevertheless, it has been demonstrated that rainfall intensity will increase, causing more erosion and soil and nutrient losses. In our division, a study has measured the effects of different agricultural practices on erosion and runoff under a future rainfall. On a field cultivated with sugar beets, contrasted tillage practices were tested. Simplified tillage with decompaction produces less soil losses, while winter ploughing gives the lowest runoff quantities comparing with fall ploughing and soil decompaction. For a corn crop, distributed seeding (obtained with a seeder) gives the lowest rates of both soil losses and runoff quantities, comparing with classic seeding (75-cm interrow) and classic seeding with Ray-grass seeding between the rows (Kummert et al., 2011).

As we can see, modelling is really useful to represent systems and predict future conditions. But it is also important to perform field measurements in order to better understand and quantify processes. That is why we are also monitoring the spatial distribution of erosion and deposition on a pilot catchment. Instrumentation includes a weather station with disdrometer and discharge measurement at the outlet coupled with water sampling. Field observations are done to determine the texture redistribution. Moreover, very accurate digital elevation models will be obtained regularly using Lidar technology, which will allow us to follow the evolution of the topography. Then, we aim at modelling both erosion and sedimentation on this catchment, and estimating net erosive flows, taking climate change into consideration. This will help us find strategies to limit erosion and runoff from their source (Pineux & Degré, 2012a).

Our research answers concrete needs from managers of rural areas and rivers. The implementation of European directives in the environmental field and, especially, in water management, generates a request from policy makers for new tools. Flood maps taking into account overland flows and mudflows in addition to river overflowing have been produced as part of the implementation of the Floods Directive in Wallonia. Indeed, on top of loss of arable land, runoff can cause great damage to property and deteriorate water quality in watercourses (Pineux & Degré, 2012b). EPICgrid was used to evaluate diffuse pollution and the efficiency of mitigation measures in the context of the Nitrate Directive. The knowledge of river behaviour during low flows is also important nowadays for a good integrated management of rivers (managing water quality in regards to the Water Framework Directive, biodiversity, water resources, etc.). Another study has allowed water managers to calculate low flow discharges (MAM7 i.e. Mean Annual Minimum flow on a 7-day average basis) in any ungauged catchments of Wallonia for different return periods (Grandry et al., 2012).

In conclusion, all these natural phenomena (floods, drought, erosion, mudflows, nitrate leaching, etc.) will be affected by climate change and especially by extreme conditions. But they can also be influenced by humans. We need to adapt, and measures have to be taken in order to reduce negative effects of climate evolution and take the best of the potential positive effects. We believe that adapting agriculture can play a major role in mitigation at plot and catchment scales.

REFERENCES

Bauwens, A., Sohier, C. & Degré, A. 2011. http://hdl.handle.net/2268/91848.
Drogue, G. et al., 2010. http://hdl.handle.net/2268/68472.
Grandry, M., Gailliez, S., Sohier, C., Verstraete, A. & Degré, A. 2012. A method for low flow estimation at ungauged sites, case study in Wallonia (Belgium). *Hydrol. Earth Syst. Sci. Discuss.* 9:11583–11614.
Kummert, N., Beckers, E. & Degré, A. 2011. http://hdl.handle.net/2268/83320.
Pineux, N. & Degré, A. 2012a. http://hdl.handle.net/2268/128148.
Pineux, N. & Degré, A. 2012b. http://hdl.handle.net/2268/120911.
Sohier, C., Degré, A. & Dautrebande, S. 2009. http://hdl.handle.net/2268/16572.
Sohier C. 2011. http://hdl.handle.net/2268/86912.

Taming the rivers? Solutions and strategies

R. Thépot
Seine Grands Lacs, Paris, France

L'EPTB Seine Grands Lacs

L'Institution interdépartementale des barrages-réservoirs du bassin de la Seine est un établissement public interdépartemental qui regroupe Paris, les Hauts-de-Seine, la Seine-Saint-Denis et le Val-de-Marne.

Depuis sa création en 1969, elle a la charge d'une double mission:

– soutenir l'étiage pour maintenir les débits de la Seine et de ses affluents,
– lutter contre les risques liés aux inondations.

A cet effet, elle exploite 3 ouvrages situés en dérivation de la Seine, de la Marne, de l'Aube et un sur l'Yonne, capables de stoker plus de 800 millions de m3 d'eau.

Depuis sa reconnaissance comme Établissement public territorial de bassin en 2011, l'Institution appelée dorénavant EPTB Seine Grands Lacs voit ses actions s'élargir progressivement avec notamment comme objectif de faciliter la gestion équilibrée de la ressource en eau à l'échelle du bassin de la Seine en amont de la confluence avec l'Oise. Cette évolution implique notamment le portage d'un plan global d'action de prévention des inondations centré sur la culture du risque et comprenant la réalisation d'un casier pilote sur le site de la Bassée en Seine-et-Marne en vue de compléter le dispositif existant.

L'Intervenant

Régis Thépot est directeur général de l'EPTB Seine Grands Lacs depuis 2009. Il a précédemment occupé ce même poste au sein de l'Etablissement public Loire pendant 13 ans. Il a été par ailleurs délégué général de l'association française des EPTB de 1999 à 2005 et de 2011 à 2012.

Avec une expérience de 30 ans dans les domaines de l'eau et de l'aménagement, il intervient régulièrement dans le cadre de partenariats entre l'Europe et la Chine et de forum internationaux. Il est notamment intervenu récemment en novembre 2012 à Rotterdam à la conférence Room for the river, organisée par le Rijkswaterstaat.

The EPTB Seine Grands Lacs

Seine Grands Lacs is an interdepartmental public organization which covers Paris, Hauts-de-Seine, Seine-Saint-Denis and Val-de-Marne.

Since its creation in 1969, it is entrusted with two missions:

– low-water enrichment for the Seine and its tributaries,
– prevent flood-related risks.

To this end, it manages 3 reservoirs located on side channels of the Seine, Marne, Aube and Yonne, representing a storage of 800 millions m3 water.

In 2011, it was labeled *Établissement public territorial de bassin* (EPTB) and was given its current name EPTB Seine Grands Lacs. Its actions are being widened with, for instance, the objective to facilitate a balances management of water resources at the scale of the Seine basin upstream of its confluence with Oise. This evolution implies the leadership on the global action plan for flood prevention, centered on risk awareness and implementation of the pilot storage on the Bassée area in Seine-et-Marne to complement the existing infrastructures.

The Speaker

Régis Thépot is director general of the EPTB Seine Grands Lacs since 2009. He formerly held the same position at the Etablissement Public Loire for 13 years. He was deputy general of the association of EPTB from 1999 to 2005 and from 2011 to 2012. He has a 30-years experience in water and planning and is regularly involved with partnerships between Europe and China as well as international forums. He recently spoke at the Rotterdam conference of Room for the river in November 2012, organized by Rijkswaterstaat.

Transboundary Water Management in a Changing Climate – Dewals & Fournier (Eds)
© 2013 Taylor & Francis Group, London, ISBN 978-1-138-00039-1

How to communicate about climate change impacts?

E. Huyghe [West – Vlaamse Intercommunale (wvi)]
West Flanders Intermunicipal Association, Service Rendering Public Body, Brugge, Belgium

ABSTRACT: How can we communicate about climate change impacts and the adaptation measures necessary to cope with it? What can go wrong in the communication? And how can we prevent that our message is misunderstood or neglected?

1 INTRODUCTION

This paper looks into how the communication model developed by Roman Jakobson can be used in the communication about climate change adaptation. It also gives some recommendations and examples from the North West Europe projects SIC Adapt! and Future Cities. It does not have the purpose to give solutions with regard to communication about climate change and its impacts.

2 COMMUNICATION MODEL

The communication model (Figure 1) developed by Roman Jakobson (° 11/10/1896 – + 18/07/1982) is a very commonly used model that structures all kinds of communication.

The model consists of six elements: the sender, the message, the receiver, the code, the channel and the context.

- The sender is the person who sends a message, the person who wants to communicate.
- The message is what the sender wants to communicate, the topic.
- The receiver is the person who receives the message.
- The message is sent in a certain code.
- The sender uses a channel to get the coded message to the receiver.
- The context is about the own vision, the own world of sender and receiver.

This model seems simple but quite a lot can go wrong. This is called "noise" or "interference". What happens if the receiver cannot decode the code used by the sender? Or when the receiver is

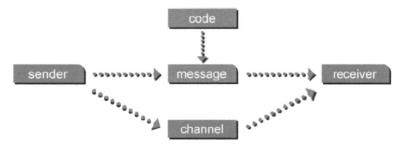

Figure 1. Communication model showing the relations between the elements sender, message, receiver, code and channel.

not able to use the channel the sender used? Or when the message is too complex? Or when the contexts of sender and receiver differ too much?

3   THE COMMUNICATION ELEMENTS WITH REGARD TO CLIMATE CHANGE IMPACTS

The sender is the person who works on climate change, its impacts and adaptation to these impacts. There is already the first danger: the sender prepares his/her message from the own context that is not the same context as the one of the receiver. A civil servant in the sustainability department lives and works in a different context than e.g. a developer of housing projects or business parks. The sender has to take into account the context of the receiver otherwise his message will not be well received.

The message is the topic the sender wants to communicate about. A message needs to be clear, unambiguous and adapted to the receiver's context. The climate change message is about uncertainties and is often unclear: what will happen? It just can't be predicted. The message is also a long term message, while a lot of people a just concerned about short term issues.

The receiver is the person that the sender wants to reach with his message. In a communication plan the target groups are the receivers. The most ideal communication is the one that is tailor-made for each different receiver. A problem with climate change is that some receivers try to ignore the possible impacts of climate change because of fear or other reasons. That is why a positive and unambiguous story is necessary (see in item 3).

The message is in a certain code: language, icons, …. For the code applies the same rules as for the message. It must be clear and unambiguous, understandable for the receiver. If a sender sends a text in English to the receiver who does not know the language, the sender has chosen the wrong code. Another example is e.g. the use of maps that the receiver does not understand.

The sender uses a channel to get the coded message to the receiver. He must take into account that the receiver must be able to use the channel.

4   EXPERIENCE FROM THE EUROPEAN PROJECT FUTURE CITIES AND THE SIC ADPAPT! CLUSTER

Future Cities is a European funded project (Interreg IVB transnational program North West Europe) in which eight partners from Germany, The Netherlands, United Kingdom, France and Belgium work together on the topic of adaptation. How city structures can be adapted to the impacts of climate change is not only a question being strategically discussed between partners but also demonstrated through many pilot projects implemented at the local level. The project is in its final phase and formulated recommendations also concerning communication about climate change and adaptation.

A first recommendation is to improve communication tools through visualization. The experience from the partners is that results of climate change studies are often too abstract for the general public. Another deficit is the lack of understandable support for decision makers and planners. A few examples from the Future Cities project are the heat attention map made in Arnhem and the water game of Tiel.

The city of Arnhem wants to deal with the heat island effect, where in summer time densely built up areas – like the city center, shopping malls and industrial areas – are far hotter than the surrounding areas. Seven to eight degrees at night is no exception. By analyzing GIS – data and experiences from German universities a heat map (Figure 2a) was developed where it is clearly shown where the hot spots of the city are. From the heat map a heat attention map (Figure 2b) was made that shows where the problem areas are and where action is necessary. This map was embedded in the new structure vision of the city and received in this way and official status in planning processes.

Figure 2a.   Heat map Arnhem.   Figure 2b.   Heat attention map Arnhem.

The water game in Tiel is a multiplayer computer game in which stakeholders take up each other's roles: a developer becomes a city civil servant, the water board becomes a developer, ... In this way all stakeholders get to know each other's stakes and interests what creates a better understanding and improved cooperation.

Another recommendation is to increase the attention for climate change and its impacts in education and lifelong learning. Climate change should be included in school curricula, also in primary and secondary schools. When climate change is described in the curricula it is mostly about mitigation while adaptation is quite often ignored. Another experience is that education tools are "by-products" of e.g. European projects and do not find their way to national authorities for inclusion in school curricula. Recommendations are to create an EU priority to include climate change in teaching methods and curricula, to identify what exists, to identify gaps, to mobilize support for teacher training, to use existing networks to enhance the exchange of experience and to ease the sharing of information through e.g. a climate change knowledge hub.

And finally, it is very important to express a positive, unambiguous and short message. The public perceives climate change impacts in a negative way through messages in the media about flooding, drought, sea level rise, ... Perhaps the adaptation message can be one of positive cooperation between different departments making the city/region more lively and attractive and at the same time better adapted to climate change impacts, without even mentioning the latter, as this has a negative connotation.

5   CONCLUSION

To conclude, three main recommendations concerning communication about adaptation to climate change can be formulated from existing experiences described earlier in this paper:

1) Make the climate change message less abstract through e.g. visualization tools;
2) Increase the attention for climate change and its impacts in education and lifelong learning by e.g. including it in school curricula of professional secondary schools;
3) Make the climate change message unambiguous, clear and keep it short.

6   WHO IS WVI?

The West – Vlaamse Intercommunale is a service rendering public association working for 54 municipalities and the province of West Flanders. In this working area live about 1 million people. Wvi develops and manages business parks and is also active in housing projects. Next to that, wvi provides the municipalities and the province with advice and studies concerning spatial and urban planning, mobility and environment and nature.

Transboundary Water Management in a Changing Climate – Dewals & Fournier (Eds)
© 2013 Taylor & Francis Group, London, ISBN 978-1-138-00039-1

La Commission Internationale de la Meuse

J. Tack
Président CIM, Liège, Belgique

1 LA COMMISSION INTERNATIONALE DE LA MEUSE

La Commission internationale de la Meuse est composée des huit Parties contractantes à l'Accord international sur la Meuse (Gand, 2002) à savoir: l'Allemagne, la Belgique, la Région de Bruxelles-Capitale, la Région flamande, la Région wallonne, la France, le Grand-duché du Luxembourg et les Pays-Bas.

Des organisations (non gouvernementales) ont un statut d'observateur auprès de la Commission et peuvent participer à certains travaux.

La Commission internationale de la Meuse dispose d'un secrétariat permanent établi à Liège, au Palais des Congrès. Le secrétariat assiste la Commission dans l'exécution de ses travaux.

2 OBJECTIFS DE LA CIM

Comme indiqué dans l'article 2 de l'Accord de Gand, les Parties contractantes s'efforcent de réaliser une gestion de l'eau durable et intégrée pour le district hydrographique international de la Meuse, compte tenu en particulier de la multifonctionnalité de ses eaux.

Elles coopèrent plus particulièrement afin de:

- coordonner la mise en œuvre des exigences définies dans la Directive cadre sur l'eau pour
- réaliser ses objectifs environnementaux et en particulier tous les programmes de mesures, pour le district hydrographique international de la Meuse;
- produire un seul plan de gestion pour le district hydrographique international de la Meuse conformément à la Directive cadre sur l'eau;
- se concerter puis coordonner les mesures pour une prévention et une protection contre les inondations compte tenu des aspects écologiques, de l'aménagement du territoire, de la gestion de la nature ainsi que d'autres domaines tels que l'agriculture, la sylviculture et l'urbanisation, et contribuer à atténuer les effets des inondations et des sécheresses y compris les mesures préventives;
- coordonner les mesures de prévention et de lutte contre les pollutions accidentelles des eaux et assurer la transmission des informations nécessaires.

3 LE PRESIDENT DE LA COMMISSION

Le président pour la période 2013–2015 est Monsieur Jurgen Tack (Flamand), administrateur général de l'INBO (Institut pour la recherche sur la Nature et les Forêts). Le Président de la

Commission est désigné par la Partie Contractante assurant la présidence, il n'intervient pas au nom de sa délégation. Le Président dirige les réunions de l'Assemblée plénière et des chefs de délégation. Il supervise en outre le secrétariat au nom de la Commission. Le Président représente la Commission, il en défend les intérêts. La présidence de la Commission est assurée pour une durée de deux ans par chaque partie contractante selon un ordre établi.

Jointly adapting to climate change in transboundary basins: the programme of pilot projects under the UNECE Water Convention

Sonja Koeppel

United Nations Economic Commission for Europe (UNECE), Secretariat of the Convention on the Protection and Use of Transboundary Watercourses and International Lakes

ABSTRACT: Cooperation on adaptation in transboundary basins is necessary and beneficial, but still rare. The programme of pilot projects and platform for exchanging experience on adaptation in transboundary basins under the UNECE Water Convention aims to support dialogue and cooperation on the design of an adaptation strategy in the transboundary context. After about 2 years of implementation of the currently 8 pilot projects some lessons learnt can be shared.

Adaptation measures taken unilaterally in one country, especially structural measures such as dams, reservoirs or dykes can have significant negative effects on other riparian countries. On the other hand, transboundary cooperation on adaptation helps to find better and more cost-effective solutions, by considering a larger geographical area in planning measures, by broadening the information base, by sharing costs and benefits of adaptation measures and by combining efforts. Although many countries are now starting to assess climate change impacts and to develop adaptation strategies for their own territory, still very little is done at the transboundary level. The AMICE project has been a pioneering initiative in this regard.

This issue is addressed by the programme of pilot projects on adaptation to climate change in transboundary basins under the United Nations Economic Commission for Europe (UNECE) Convention on the Protection and Use of Transboundary Watercourses and International Lakes (Water Convention) and its Task Force on Water and Climate. The programme aims to support countries in their efforts to develop adaptation strategies and measures in transboundary basins and to create positive examples demonstrating the benefits of and possible mechanisms for transboundary cooperation in adaptation planning and implementation. The programme includes projects, which are directly supported by the UNECE secretariat in cooperation with partner organizations in the framework of the Environment and Security Initiative (ENVSEC) – such as the projects on the Chu Talas, Dniester, Neman and Sava – as well as other on-going initiatives with their own implementing framework on the Danube, Rhine and Dauria. The AMICE project has joined the programme of pilot projects for sharing experiences.

In all the pilot projects a transboundary climate change impact and vulnerability assessment has been developed. The "Pilot project on river basin management and climate change adaptation in the Neman River Basin" has resulted in a first joint assessment of water resources and climate change impacts in the Neman Basin, thereby enabling a renewal of cooperation between the riparian countries on the shared river basin (Belarus, Lithuania, and Russian Federation). In the project "Reducing vulnerability to extreme floods and climate change in the Dniester River Basin" (Ukraine and Republic of Moldova) a first basin-wide impact and vulnerability assessment has been developed, as well as detailed flood risk modelling in two priority sites.

The programme of pilot projects which was started in 2010, also includes a platform for exchanging experience through regular meetings as well as a web-based platform. Annual workshops are organized which are open to basins worldwide and a core group of representatives of the pilot projects was created which meets annually and enables a direct exchange of experience between the projects.

The pilot projects programme is based on a sound and common basis: the Guidance on Water and Adaptation to Climate Change, which was developed by more than 80 experts, under the leadership of the Netherlands and Germany and adopted at the fifth session of the Meeting of the Parties to the Water Convention in 2009.

Some lessons learnt from the pilot projects:

- In most basins some climate change impact assessments had already been developed nationally, but using different methodologies and often not specifically for a particular river basin. This underlines the importance of developing joint scenarios, modelling and vulnerability assessment by all basin countries at the basin level. If all riparian countries already use their own models, harmonization at least at the borders should be discussed. Local experts should be involved as much as possible.

- Often, many more activities on water and climate change have already been performed in the basins than expected. Therefore, it is important to start any project with a thorough baseline study and to establish links with all relevant actors such as local and national authorities, academia, NGOs, relevant business and international organizations.

- Connecting research and policy-making, bringing together experts/scientists and decision-makers is crucial from the beginning, even in the project elaboration phase in order to ensure ownership. This can be done through the creation of a permanent working group with all different stakeholders represented, like in the Dniester pilot project where decision-makers were involved in the selection of sites for flood modelling.

- Institutional and cultural differences may complicate transboundary cooperation but these can be overcome. For example, countries can look for what they have in common as a basis for cooperation, like a shared interest. The project 'Dauria going dry', for instance, implemented in a region where competition over water makes countries hesitant to share information, focuses on nature conservation and climate change adaptation which all countries are interested in for various reasons.

- In transboundary basins, ensuring coordination and cooperation between adaptation at different governance levels (from local to transboundary level) is crucial.

- It is important to combine climate change impact research and strategy development with implementation of some priority no- or low-regret measures such as installation of monitoring stations, small-scale restoration of ecosystems or awareness-raising activities. The AMICE project provides a very good model for this.

The pilot projects programme under the UNECE Water Convention is now transformed into a global network of basins working on climate change adaptation, in cooperation with the International Network of Basin Organizations (INBO). In 2013–2015, the exchange of experience through annual workshops will continue and a collection of good practices will be elaborated. Other basins and projects are welcome to join.

FURTHER INFORMATION

Project details are available at <http://www.unece.org/env/water/water_climate_activ.htm>.
And at: http://www1.unece.org/ehlm/platform/display/ClimateChange/Welcome.
UNECE. 2009. *Guidance on Water and Adaptation to Climate Change*. Available at http://www.unece.org/env/water/publications/documents/Guidance_water_climate.pdf.

AMICE song

J. De Bijl
Waterboard Aa en Maas, 's-Hertogenbosch, The Netherlands

M. Linsen
Rijkswaterstaat, Lelystad, The Netherlands

The Meuse and his affluents flow calm sometimes wild
From France to the North-sea meandering around
She's connecting the people the animals and plants
In these beautiful surrounding inspiration is found

Refrain:
So for now and forever (handclap 4 times) Amice will sustain
And our co-o-peration will always remain

Due to global warming the climate will change
The discharge of the river will severely derange
Serious droughts and the rai-ain will pour
So we are told by the famous Al Gore

(Refrain)

If we work together on a river basin scale
A sustainable and secure water future won't fail
No borders one river and all share one goal
United we stand and divided we fall

(Refrain)

A transnational roadmap for adaptation is found
Measures for low flows and floods all around
Knowledge and experience between partners exchanged
And the people of the Meuse-basin are all engaged

(Refrain)

New challenges in future coming up in Meuseland
Integration with the frame work is already planned
Together we'll search for the measures that match
And the Meuse will be adapted to climate impacts

(Refrain)

Old river Meuse meets Super hero AMICE

M. Lejeune
RIOU, Hasselt, Belgium

M. Fournier
EPAMA, Charleville-Mézières, France

ABSTRACT: This dialogue was designed to present the results of the AMICE Project in an original way during the Interreg IV B Programme's annual event, Dortmund, 2012. One speaker plays the Meuse river while the second speaker plays the AMICE part. Costumes were added to make it more enjoyable. Questions were also asked to the audience about what they would have done to help the Meuse. The answers were compared with the AMICE solutions. The dialogue and interactions with the audience were moderated by Jodie Bricout from the Capem-project.

1 SOMETHING IS WRONG WITH THE RIVER MEUSE

Hello, who are you?

River Meuse: I'm old, old river Meuse. I've been around for a long time and I've seen it all: Ice Ages, mountains rising and going down, parts of my basin being stolen away by hostile rivers, being forced to cut my way through the hard Ardennes mountains using my power and my strength and showing off my stubborn character. And people! You've used me for all kinds of things: transportation, migration, waging wars, for taking your garbage, for cooling water, drinking water (for 9 million people) and so on. You even invented things like countries with me as a frontier! Being considered as a frontier by people was and still is the greatest humiliation. As if I want to divide anything, me a river, who only wants to connect: upstream and downstream, left bank and right bank.

People love me though, at least when I'm being good, which means when I have a nice, average $250\,m^3/s$ discharge. But like many old things I can be rather eccentric, even cantankerous and I can change myself into a raging fury with discharges reaching up to more than $3000\,m^3/s$. Then people think they can say I'm being nasty! They say they don't want all the water! But then later, when I relax a bit and take a break, which means there is hardly any water in my riverbed, then they all want the water, even fighting over it! Ha ha, but when I brought it, they didn't want it.

And now there is climate change coming along! Listen, I think I've seen worse in my long long life but still it frightens me.

Aren't there people helping you?

River Meuse: Oh yes they are, or rather they think they are. But while they think they are helping me, they all act only in their own country not even looking at the upstream or downstream areas. As if the water stops at those man-made frontiers! As if I'm a sausage that can be sliced up! I'm a dynamic river, not a static road.

Moreover all these humans 'helping' me don't even know me properly. Ask in France: they'll say the Meuse, that's somewhere in the Nord-East isn't it? Ask in the Netherlands, they think I'm a Dutch river with a small tail in Belgium and France. How can they help me when they don't understand me?

AMICE: Ah ah ah but now AMICE is there!
River Meuse: AMICE?
AMICE: yes, and I want you to become the very best example of a climate proof river.
River Meuse: I wonder how you will do that! I really have some big problems, you know.
AMICE: I am a super hero you know! I will solve your problems, be patient and you will see.

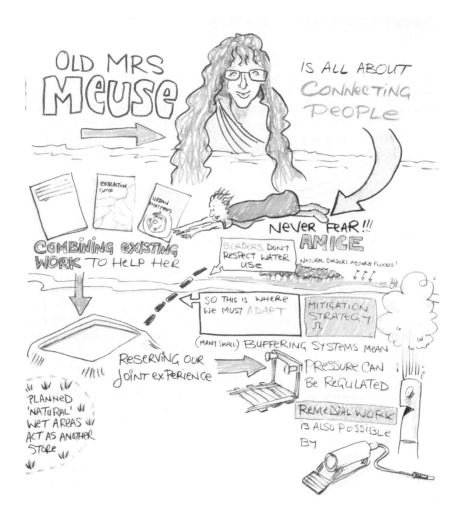

2 AMICE FINDS OUT WHAT IS WRONG WITH THE MEUSE AND WHAT SHOULD BE DONE

So, river Meuse, what's your problem?

River Meuse: My problem? There's more than just one! First, what's this climate change? There are a whole bunch of scenarios: an IPCC one and, of course, a Dutch, French, Belgian and German one. And no, they don't all point in the same direction! How do you think I can discharge more water and less water at the same time? How can I possibly know what's going to happen?

AMICE: Well AMICE is very powerful so it can integrate several scenarios and make one that fits just well with the Meuse basin. Oh! Oh! But the future scenarios warn that you will become more lunatic … old rivers tend to have a worsening temper.

River Meuse: Everybody says there will be more heavy rains! And of course, every country had its own version of hydraulic modeling to tell me what to do; but let me tell you a secret? Well, according to these models, my water was jumping up or even disappearing at the countries' borders! As if THAT could be possible!

AMICE: Well AMICE is very clever so it has found a way to paste the models together and now I can tell you exactly where the water goes and how deep it is in any single point of you.

River Meuse: With all these heavy rains coming, it's so obvious that I cannot possibly keep all the water within my banks. Will I have to spill it everywhere, flooding towns and villages and making people hate me again?

AMICE: No, no, stop crying! There are several better solutions for this.

- We build reservoirs for temporary storage of the extra water
- We give you more room within your banks so that you don't have to spill over
- We enhance the natural capacity for water retention of the tributaries' river valleys so more water can be held there.

Well AMICE can do everything! I will make a little reservoir here, and enlarge the banks there and also improve the natural valleys all over here. And houra, problem solved!

River Meuse: And then after the flood is gone I might have a tantrum and want some rest, because I'm old and that's what old people do. Then there'll be a drought. And then, of course, everybody wants to take water out of me! And no, they don't think about their neighbors.

AMICE: You eccentric old river you! Of course we have solutions for this too.

- Prioritized users can get all the water they need: the more important you are, the more water you can use
- The water is used and reused before it's allowed to flow into the sea
- Thanks to the reservoirs and wetlands there is a sufficient water supply to see us through the dry period

Well AMICE likes to experiment new stuff. So you will get a bit of all this, river Meuse! If people still complain after that, it's because you have decided to be really annoying.

3 THE MEUSE IS HAPPY NOW

So what about high waters and floods? Can you tell me more?

AMICE: I designed a couple of smaller projects to enhance the natural water storage capacity. Of course this is not enough, but the idea is that they can serve as sources of inspiration for other people in other regions. Our goal would be: adaptation everywhere!

River Meuse: I get it! Show me what you did.

AMICE: In the village of Ny, in the Belgian province of Luxemburg, a relatively small investment will improve the resilience of the village. An embankment has been constructed to retain superfluous water during heavy rainfall.

In the Amblève sub-basin, the natural hydrology of the upstream part of some small tributaries is being restored. The idea is that healthy ecosystems are more resistant to climate extremes, because upstream ecosystems have a 'sponge' effect, they hold water when it falls and release it slowly in times of drought.

The Steenbergsche Vliet is a small river in the very downstream part of the Meuse basin. It flows into the Volkerak-Zoom lake. In the future, the Volkerak-Zoom will be used for water storage during high volume flows of the river Meuse and her brother the Rhine. The discharge of the Steenbergse Vliet into the Krammer-Volkerak will consequently be blocked and this will result in (too) high water levels in the upstream areas. New water retention areas are being created which will support multifunctional land use (water storage, recreation, housing in flood plains, new role for historical heritage).

River Meuse: And what about drought and water scarcity?

AMICE: well in fact drought and floods are two sides of the same problem: if the water is stored in a smart way when there is plenty, there will be fewer problems when there is a drought. Thanks to AMICE, coordinated measures are applied all over the basin. And as well, here's a very nice example of a project where water is used and reused again before it is allowed to flow into the sea.

The Albert canal runs from the inland port of Liège to the port of Antwerp and is fed entirely by the Meuse. There are six locks to manage navigation over the total fall of 55 meters. In dry periods there is not enough water for effective lock function. Pumping back the lock water is a good way to save water but it has to happen without adverse effects on fish migration. In the Albert canal, this is done using huge Archimedes screws. This leads to enhanced fish migration, hydropower (when water is abundant), and the increased availability of water during droughts.

In another country, existing huge reservoirs are being adapted for the expected climate changes. Along the German river Rur there is a complex system of dams. They can be used for discharge control, to serve the purposes of drinking water, recreation, (paper) industry, hydropower and wildlife. Changing rainfall patterns pose a challenge to the guaranteed discharges during droughts and to the buffer capacity during the "floods season". These two situations could be in conflict, but with well-defined operational procedures, an optimum may be found.

River Meuse: And what about towns?

AMICE: Here's a good example and one which should be inspirational for other towns. It is the city of 's Hertogenbosch in The Netherlands.

The tributaries of the Meuse can have peak discharges simultaneously. Measures in the main river upstream lead to the possibility of a higher peak in a shorter period near 's-Hertogenbosch. An area of 750 hectares will be adapted to become a water retention area. It is designed to be flooded once every century. The peak discharges can be saved, and the water has room to continue its course.

People can also learn to react to an extreme event and use computer-based tools to help them! Flood crisis exercises are organised to train the emergency services and to test how the information is shared. Last year I had 300 people running around to fight a flood! But there was no water in the streets of course, it was just an exercise. They learned a lot!

Due to climate change next summer could be a hot one. But the Meuse can cope now! The water is safely stocked and look how beautiful the sandy beaches along the natural banks are!

River Meuse: At last I'm treated as a river now, with proper respect.

But wait for my next surprises!

Transboundary Water Management in a Changing Climate – Dewals & Fournier (Eds)
© 2013 Taylor & Francis Group, London, ISBN 978-1-138-00039-1

Internet documentary on the river Meuse
www.amice-film.eu

W. Overmars
Rombus Natuurfilms, The Netherlands

The Amice film tells the story of the Meuse as the river and its tributaries flow through the river basin. The water travels through very different landscapes. In France, it flows as a small river in a wide valley with often ample floodplains. In Wallonia the river is surrounded by the low mountains of the Ardennes. Here, the river receives most of its water from its tributaries the Semois, Sambre, Ourthe, Lesse, and the Vesdre. Past the city of Maastricht, the Meuse is over a length of 50 km the border between the Netherlands and Flanders. Luckily, these days the river is considered to be a shared richness and is called the 'common' Meuse. The last large tributary, the Rur, brings the water from the German Eifel hills. In the Netherlands, the river enters the low country and the waters mingle with those of the Rhine. In the wide arms of the Rhine-Meuse estuary the river disappears into the see.

During the early middle ages, the river basin as a whole was an important cultural and economic unity, the heartland of the Carolingian Empire. In the disturbances of later ages, the river was cut in parts: a French, Belgian, German and a Dutch one, treated as different rivers by people who knew too little about each other.

In the AMICE-film the river is restored into its earlier coherence. The common history and culture is emphasized, as well as the hydrological unity of the system. The many commodities the river offers us, from cooling water for nuclear plants to the drinking water for millions of people as well as a rich ecosystem are shown. Common problems of the river, like the danger of floods, the problems of drought or the uncertainties of climate change are shared by the AMICE-partners, and reflected in the internet documentary. The AMICE-film is not an enumeration of the AMICE achievement. Instead, it is a contribution in itself to restore a common consciousness of sharing a beautiful, rich, useful, safe and healthy catchment area.

Transboundary Water Management in a Changing Climate – Dewals & Fournier (Eds)
© 2013 Taylor & Francis Group, London, ISBN 978-1-138-00039-1

AMICE Partners

LEAD PARTNER

1. EPAMA
   26, avenue Jean Jaurès – 08000 Charleville-Mézières – FRANCE

PROJECT PARTNERS

2. University of Lorraine – Centre d'Etudes Géographiques
   (CEGUM)
   Ile du Saulcy – BP 794 – 57012 METZ CEDEX – FRANCE

3. CETMEF
   2, boulevard Gambetta – BP 60 039 – 60321 Compiègne
   cedex, FRANCE

4. Région Wallonne – GTI
   SPW – DGO 2 – DO 223 – GTI, rez de chaussée,
   Boulevard du Nord, 8, 5000 Namur, BELGIUM

Wallonie

5. University of Liege (ULg)
   Département ArGEnCo- MS^{2}F
   Hydraulics in Civil and Environmental Engineering (HECE)
   Chemin des Chevreuils, 1, 4000 – Liege, BELGIUM
   Aquapôle
   Chemin des Chevreuils, 3, 4000 – Liege, BELGIUM

6. Gembloux AgroBioTech (Gx-ABT)
   Département Hydrologie & Hydraulique Agricole,
   Passage des Déportés, 2 – 5030 Gembloux, BELGIUM

7. Agence de Prévention et Sécurité (APS)
   Rue de la Plaine, 11, 6900 Marche-en-Famenne, BELGIUM

8. City of Hotton
   Administration communale, Rue des Ecoles,
   50 – B 6990 Hotton, BELGIUM

9.   nv De Scheepvaart
     Havenstraat 44, 3500 Hasselt, BELGIUM

10.   Flanders Hydraulics Research (FHR)
     Berchemlei 115, B-2140 Borgerhout, BELGIUM

11.   RIOU asbl
     Andreas Vesaliuslaan 8, 3500 Hasselt, BELGIUM

12.   Wasserverband Eifel-Rur (WVER)
     Eisenbahnstraße 5, 52353 Düren, GERMANY

13.   RWTH Aachen University – IWW
     Mies-van-der-Rohe-Straße 1, 52056 Aachen,
     GERMANY

14.   LFI-RWTH – Aachen
     Mies-van-der-Rohe-Straße 1, 52056 Aachen,
     GERMANY

15.   Rijkswaterstaat (RWS)
     Postbus 17, 8200 AA, Lelystad,
     THE NETHERLANDS

16.   Waterboard Aa en Maas
     Pettelaarpark 70, 5216 PP, 's-Hertogenbosch,
     THE NETHERLANDS

17.   Waterboard Brabantse Delta
     Hof van Bouvigne, Bouvignelaan 5,
     4836 AA Breda, THE NETHERLANDS

Author index

T - #0785 - 101024 - C0 - 246/174/6 [8] - CB - 9781138000391 - Gloss Lamination